With 100 events in 25 countries, and 2020 is looking like a good year for Maker Faire

── JOIN OUR LINEUP TODAY! ──

JANUARY:
- ★ Bangkok
- ★ Bonita Springs

FEBRUARY:
- ★ Cairo
- ★ Kuwait
- ★ Caguas
- ★ Dubai
- ★ Elkhorn
- ★ Roanoke
- ★ Jacksonville

MARCH:
- ★ Pilsen
- ★ Rhur
- ★ Sachsen
- ★ Malta
- ★ The Big Muddy
- ★ Raritan Valley
- ★ Middlesex County
- ★ Lafayette
- ★ Cache Valley
- ★ Lynchburg
- ★ Greater Lafayette

APRIL:
- ★ Zadar
- ★ Lille
- ★ Berlin
- ★ Edinburgh
- ★ Hsinchu
- ★ Westport
- ★ Miami
- ★ Meridian
- ★ Pioneer Valley
- ★ Wichita
- ★ Martinsville
- ★ Cape Cod
- ★ Tyler

- ★ Loma Linda
- ★ Mercer-Bucks
- ★ Asheville
- ★ Bowling Green
- ★ Cranberry Township
- ★ South Bend
- ★ Bloomsburg

MAY:
- ★ Vienna
- ★ Saskatoon
- ★ Dugo Selo
- ★ Kyoto
- ★ Luxembourg
- ★ Alentejo
- ★ Burlington
- ★ Gilroy
- ★ North Little Rock
- ★ Columbia
- ★ Hawthorne
- ★ Stark

JUNE:
- ★ Prague
- ★ Trieste
- ★ Manila
- ★ Glasgow
- ★ Long Island
- ★ South Shore
- ★ Sheboygan
- ★ Minneapolis-St. Paul
- ★ NoVA

JULY:
- ★ Edmonton
- ★ Windhoek
- ★ Detroit
- ★ Coeur D'Alene
- ★ Kingsport
- ★ Tampa

AUGUST:
- ★ Tokyo
- ★ Seoul

SEPTEMBER:
- ★ Hannover
- ★ Lisbon
- ★ Moscow
- ★ Eindhoven
- ★ Tulsa
- ★ Milwaukee
- ★ Louisville
- ★ Atlanta
- ★ Des Moines
- ★ Shreveport-Bossier
- ★ Rogue Valley

OCTOBER:
- ★ Zagreb
- ★ Rome
- ★ Warsaw
- ★ The Ozarks
- ★ Downtown Columbia
- ★ Fredonia
- ★ Philadelphia

NOVEMBER:
- ★ Paris
- ★ Sindelfingen
- ★ Jalisco
- ★ Taipei
- ★ Cleveland
- ★ Orlando
- ★ Rochester
- ★ Pensacola
- ★ York County
- ★ Baton Rouge
- ★ Madison

makerfaire.com

20

ON THE COVER:
FIX OUR PLANET — our appeal to makers everywhere to help lead the charge of arresting global warming before it's too late. Read how on page 24.

Artwork: Main image – Lifeking, TAW4, and Yevhenii via Adobe Stock. Jimmy DiResta – Mark Adams

TASTY TECH

30

38

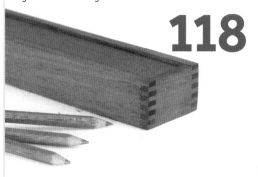

STATEMENT OF OWNERSHIP, MANAGEMENT AND CIRCULATION (required by Act of August 12, 1970: Section 3685, Title 39, United States Code). 1. MAKE Magazine 2. (ISSN: 1556-2336) 3. Filing date: 10/1/2019. 4. Issue frequency: Quarterly. 5. Number of issues published annually:4. 6. The annual subscription price is 34.95. 7. Complete mailing address of known office of publication: Make Community, LLC 150 Todd Road Ste. 200, Santa Rosa, CA 95407. Contact person: Kolin Rankin. Telephone: 305-859-0063 8. Complete mailing address of headquarters or general business office of publisher: Make Community, LLC 150 Todd Road Ste. 200, Santa Rosa, CA 95407. 9. Full names and complete mailing addresses of publisher, editor, and managing editor. Publisher, Todd Sotkiewicz, Make Community, LLC, 150 Todd Road Ste. 200, Editor, Mike Senese, Make Community, LLC, 150 Todd Road Ste. 200, Santa Rosa, CA 95407, Managing Editor, N/A, Make Community, LLC, 150 Todd Road Ste. 200, Santa Rosa, CA 95407. 10. Owner: Make Community, LLC; 150 Todd Road Ste. 200, Santa Rosa, CA 95407. 11. Known bondholders, mortgages, and other security holders owning or holding 1 percent of more of total amount of bonds, mortgages or other securities: None. 12. Tax status: Has Not Changed During Preceding 12 Months. 13. Publisher title: MAKE Magazine. 14. Issue date for circulation data below: Oct/Nov 2019. 15. The extent and nature of circulation: A. Total number of copies printed (Net press run). Average number of copies each issue during preceding 12 months:81,540. Actual number of copies of single issue published nearest to filing date: 74,128. B. Paid circulation. 1. Mailed outside-county paid subscriptions. Average number of copies each issue during the preceding 12 months: 54,817. Actual number of copies of single issue published nearest to filing date: 52,403. 2. Mailed in-county paid subscriptions. Average number of copies each issue during the preceding 12 months: 0. Actual number of copies of single issue published nearest to filing date:0. 3. Sales through dealers and carriers, street vendors and counter sales. Average number of copies each issue during the preceding 12 months: 6,256. Actual number of copies of single issue published nearest to filing date: 5,110. 4. Paid distribution through other classes mailed through the USPS. Average number of copies each issue during the preceding 12 months: 0. Actual number of copies of single issue published nearest to filing date: 0. C. Total paid distribution. Average number of copies each issue during preceding 12 months: 61,073. Actual number of copies of single issue published nearest to filing date:57,513. D. Free or nominal rate distribution (by mail and outside mail). 1. Free or nominal Outside-County. Average number of copies each issue during the preceding 12 months:722. Number of copies of single issue published nearest to filing date: 752. 2. Free or nominal rate in-county copies. Average number of copies each issue during the preceding 12 months: 0. Number of copies of single issue published nearest to filing date: 0. 3. Free or nominal rate copies mailed at other Classes through the USPS. Average number of copies each issue during preceding 12 months: 0. Number of copies of single issue published nearest to filing date: 0. 4. Free or nominal rate distribution outside the mail. Average number of copies each issue during preceding 12 months: 1,469. Number of copies of single issue published nearest to filing date: 778. E. Total free or nominal rate distribution. Average number of copies each issue during preceding 12 months: 2,190. Actual number of copies of single issue published nearest to filing date: 1,530. F. Total free distribution (sum of 15c and 15e). Average number of copies each issue during preceding 12 months: 63,263. Actual number of copies of single issue published nearest to filing date: 59,043. G. Copies not Distributed. Average number of copies each issue during preceding 12 months: 18,276. Actual number of copies of single issue published nearest to filing date: 15,085. H. Total (sum of 15f and 15g). Average number of copies each issue during preceding 12 months: 81,540. Actual number of copies of single issue published nearest to filing: 74,128. I. Percent paid. Average percent of copies paid for the preceding 12 months: 96.54% Actual percent of copies paid for the preceding 12 months: 97.41% 16. Electronic Copy Circulation: A. Paid Electronic Copies. Average number of copies each issue during preceding 12 months: 21,338. Actual number of copies of single issue published nearest to filing date: 21,951. B. Total Paid Print Copies (Line 15c) + Paid Electronic Copies (Line 16a). Average number of copies each issue during preceding 12 months: 82,411. Actual number of copies of single issue published nearest to filing date: 79,464. C. Total Print Distribution (Line 15f) + Paid Electronic Copies (Line 16a). Average number of copies each issue during preceding 12 months: 84,601. Actual number of copies of single issue published nearest to filing date: 80,941. D. Percent Paid (Both Print & Electronic Copies) (16b divided by 16c x 100). Average number of copies each issue during preceding 12 months: 97.41%. Actual number of copies of single issue published nearest to filing date: 98.11%. I certify that 50% of all distributed copies (electronic and print) are paid above nominal price: Yes. Report circulation on PS Form 3526-X worksheet 17. Publication of statement of ownership will be printed in the Dec/Jan 2020 issue of the publication. 18. Signature and title of editor, publisher, business manager, or owner: Todd Sotkiewicz, Business Manager. I certify that all information furnished on this form is true and complete. I understand that anyone who furnishes false or misleading information on this form or who omits material or information requested on the form may be subject to criminal sanction and civil actions.

Mark Adams, Aburie Pick, Rima Khalek, Debra Ansell, Dan Struffolino , Florian Hu

Make:

> "Opportunity is missed by most people because it is dressed in overalls and looks like work" — Thomas Edison

PRESIDENT
Dale Dougherty
dale@make.co

VP, PARTNERSHIPS
Todd Sotkiewicz
todd@make.co

EDITORIAL

EXECUTIVE EDITOR
Mike Senese
mike@make.co

SENIOR EDITORS
Keith Hammond
keith@make.co
Caleb Kraft
caleb@make.co

PRODUCTION MANAGER
Craig Couden

CONTRIBUTING EDITOR
William Gurstelle

CONTRIBUTING WRITERS
Debra Ansell, Brian Bunnell, John Collins, Larry Cotton, Rich Gibson, Saul Griffith, Ben Hobby, Florian Hu, Rima Khalek, Bob Knetzger, Hiroshi Maeda, Dan Maloney, Samer Najia, David Picciuto, Jonathan Stapleton, Alex Wulff

DESIGN & PHOTOGRAPHY

CREATIVE DIRECTOR
Juliann Brown

MAKE.CO

ENGINEERING MANAGER
Alicia Williams

WEB APPLICATION DEVELOPER
Rio Roth-Barreiro

GLOBAL MAKER FAIRE

MANAGING DIRECTOR, GLOBAL MAKER FAIRE
Katie D. Kunde

MAKER RELATIONS
Sianna Alcorn

GLOBAL LICENSING
Jennifer Blakeslee

MARKETING

DIRECTOR OF MARKETING
Gillian Mutti

OPERATIONS

OPERATIONS DIRECTOR
Cathy Shanahan

ACCOUNTING MANAGER
Kelly Marshall

OPERATIONS MANAGER & MAKER SHED
Rob Bullington

PUBLISHED BY

MAKE COMMUNITY, LLC
Dale Dougherty

Copyright © 2019-2020 Make Community, LLC. All rights reserved. Reproduction without permission is prohibited. Printed in the USA by Schumann Printers, Inc.

Comments may be sent to:
editor@makezine.com

Visit us online:
make.co

Follow us:
🐦 @make @makerfaire @makershed
📘 makemagazine
📷 makemagazine
▶ makemagazine
📺 twitch.tv/make
Ⓟ makemagazine

Manage your account online, including change of address:
makezine.com/account
866-289-8847 toll-free in U.S. and Canada
818-487-2037,
5 a.m.–5 p.m., PST
cs@readerservices.makezine.com

Make: Community

Support for the publication of Make: magazine is made possible in part by the members of Make: Community. Join us at make.co.

CONTRIBUTORS

When you're in the middle of a project in your shop, what's your go-to meal or snack to keep you working fast?

Jonathan Stapleton
Essex, VT
(Flat-Out Flinger)

My go-to snack is a handful of cashews — assuming there's no cold pizza in the fridge.

Alex Wulff
Skaneateles, NY
(Storm Warning)

I munch on sunflower seeds constantly while I'm making, but if I'm really hungry I'll snack on some cheese and crackers!

Debra Ansell
Los Angeles, CA
(LED "Inner Glow" Heart)

I'm not discriminating. I'll grab anything nearby that can be eaten one-handed!

Issue No. 71, December 2019/January 2020. *Make:* (ISSN 1556-2336) is published quarterly by Make Community, LLC, in the months of February, May, Aug, and Nov. Make Community is located at 150 Todd Road, Suite 200, Santa Rosa, CA 95407. SUBSCRIPTIONS: Send all subscription requests to *Make:*, P.O. Box 17046, North Hollywood, CA 91615-9588 or subscribe online at makezine.com/offer or via phone at (866) 289-8847 (U.S. and Canada); all other countries call (818) 487-2037. Subscriptions are available for $34.99 for 1 year (4 issues) in the United States; in Canada: $43.99 USD; all other countries: $49.99 USD. Periodicals Postage Paid at San Francisco, CA, and at additional mailing offices. POSTMASTER: Send address changes to *Make:*, P.O. Box 17046, North Hollywood, CA 91615-9588. Canada Post Publications Mail Agreement Number 41129568. CANADA POSTMASTER: Send address changes to: Make Community, PO Box 456, Niagara Falls, ON L2E 6V2

PRINTED WITH SOY INK

DIY for the Win

BACK FROM THE BRINK

Make: is the only print subscription that I've kept because there's something satisfying about holding and collecting the issues. I don't remember how I stumbled across the magazine, but it was several months before the first issue was published. I was not a maker at the time; indeed, the term "maker" didn't exist in its current context.

I guess what I'm trying to say is that *Make:* satisfied an urge that I didn't know existed. When Volume 01 arrived, I stayed up all night reading it from front to back and back again. With the exceptions of issues 44 and 56, my collection has survived multiple moves to multiple cities, and it is my first choice for both casual reading and focused reference. Most of all, *Make:* inspired me to think in a different way about making things. Over the last few years, I've cobbled together little DIY projects,

mostly to meet my domestic needs. They're crude and not very pretty, but they get the job done and provide the inspiration for the next iteration.

So congratulations for keeping *Make:* alive. May it inspire many future generations of makers worldwide. —*Bill Doorley, Pittsburgh, PA*

ORIGINAL SIZE

Just want to say: Yay! and Thanks!
—*Steven Lumos, via email*

MAKE ON

I just read this article ("Maker Media Has Shut Down. But Founder Dale Dougherty Isn't Calling It Quits," edsurge.com/news/2019-06-09-a-call-to-remake-the-maker-faire), so I decided to email you to simply say: thank you. When I found the world of making through Particle, Adafruit, and yes of course, *Make:* magazine, my life changed. Not dramatically or drastically like some other super inspirational stories out there, but it did. Making things gave me the creative outlet I craved in a medium I thought I would never be able to touch. It made my life better, and it's because of people like you who love to inspire people to be creative and gave this community a name and purpose to rally behind together.

So thank you, and let's keep on making.
—*Charlyn Gonda, via email*

MORE THAN WORDS

I can't tell you how much I've enjoyed my subscription, the website, THE BOOKS (mostly via Humble Bundle), and of course the Bay Area Maker Faire. —*John Bernstein, via email*

→ "Just fits" —*Steve Cooke, via email*

MAKE: AMENDS

In "**Get Nybbled!**" (*Make:* Volume 70, page 76) author Rongzhong Li released the first OpenCat demo on Maker Share in 2018, not 2016. We regret the error.

Tasty Tweets

Patrick Ohlson
@pp3dp_se

Recieving the 70th issue of #makemagazine does make me all feely @make 🖤 🍽 ⚒ #makersgonnamake

Tuesday 10/08/2019

Having trouble viewing this email? View in Browser

Welcome to Your New Issue of Make: Magazine!

Make: magazine is back in action and back to our original size! This issue's cover project is a maker's take on a Boston Dynamics-style quadrupedal walking robot that you can build yourself. Then, build an adorable unicorn-shaped dispenser that spits soap on command. And to celebrate *Make:*'s return, why not build a custom dancing version of our Makey mascot.

Plus, 28 projects including:

• Teeny-tiny personal motorboat
• Standup paddle board
• Bird-identifying computer-vision birdhouse

Volume 70
Oct/Nov 2019

Jim St. Leger
@JimStLeger

It's baaaacccckkkkk!!!!!!!
@Make Magazine has returned, and in the awesome original size too!!! Thank you @dalepd and the @MakerMedia team! You made my month! Gets yours today! #make #maker #DIY #3Dprinting #Arduino #RPi

9:41 PM · Oct 18, 2019 · Twitter for iPhone

Welcome to *Make:* Vol. 71

Hello fellow makers! We're thrilled to bring you this new issue of *Make:* magazine, the second in our throwback, original journal size. In the ensuing pages you'll find a wealth of great features, projects, and skill builders, plus a couple new regular departments: a series on YouTube makers (meet Jimmy DiResta on page 20), and profiles of project communities (get to know Hackaday.io on page 14). We also explore fun and futuristic food and beverage machines, and some unusual protein sources: check out our Tasty Technology section on page 30.

On the serious side, we also debut our 2020 focus on makers and climate change, a discussion we feel strongly about. Through our energy consumption and related emissions, humanity's environmental impact has already passed the tipping point with severe ramifications. Now, with our technological know-how, we'll need to enact radical changes to arrest the rising global temperatures that have already contributed to significant disasters, loss of life, and loss of prosperity. Saul Griffith's four-part series, beginning in this issue (page 24) will drill in on electrifying our vehicles, homes, and public infrastructures, and how makers can lead the charge. Throughout the year we'll be bringing you hands-on projects to show the way. We hope you'll give it serious thought and find ways to take part. There's a lot to do, but together, we can make it happen.

Are you doing any work or undertaking projects to help solve climate change? I'd love to hear about it, and help share your endeavors with the world. We're commited to helping with the grand project of "decarbonizing" America (and the planet). Please send me a note: mike@make.co

Regards,
Mike Senese
Executive Editor, *Make:*

MADE ON EARTH

Backyard builds from around the globe

Know a project that would be perfect for Made on Earth?
Let us know: *editor@makezine.com*

CREATIVE CUTLERY

INSTAGRAM.COM/GRIFFINKIWI

According to **Karina Olsen**, the biggest mistake you can make in welding class is telling your instructor that you're finished early. "Once the instructor knows that you need more work, they will find random, ungraded projects for you to do until your classmates catch up with you!"

When Olsen learned this lesson, she didn't consider herself a sculptor yet; however that would change as she snuck in time for experimentation. She picked up some random bits of steel at a thrift shop, typically silverware, and simply tried various ideas. It was in this way that Olsen's first sculpture was created "in secret" while she was a student in a welding class.

This first success sparked her interest in creating metal sculptures, so she scoured Milwaukee's thrift stores for utensils. Friends and family also began donating unwanted wares, allowing her to amass a library of material from which to draw inspiration.

Olsen's creations, typically animals or insects, are gorgeous feats of layered and flowing lines. The structures of the creatures themselves seem to have an intrinsic flow and she carefully uses the silverware's flourishes and scrollwork to add visual intensity to the designs.

"People don't always notice how beautiful their flatware is until it's transformed into something else," Olsen says, "and I'm always happy to help them notice by drawing attention to it in my work."

Olsen is still a student, and has a full-time job as a caretaker for the elderly, but continues to make time to sculpt and work on commissions. "There aren't many things more motivating than knowing that when I'm finished, my piece will go to a good home!"
—*Caleb Kraft*

Karina Olsen

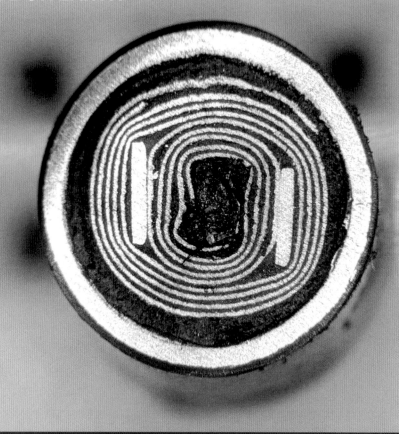

GRINDCORE TWITTER.COM/TUBETIMEUS

When your function generator dies, you never know what you'll find inside. Maybe it will be a blown tantalum capacitor, or maybe it will be an ongoing art project with followers all over the Twitterverse.

In his case, engineer **Eric Schlaepfer** found both. He wanted to see if there was an obvious failure point visible in the capacitor, so he sanded it in half to inspect, and posted it on Twitter. The feedback was overwhelmingly positive, so Eric just kept grinding away at other components.

"Some of them look quite surprising inside, and many of them look quite beautiful," Schlaepfer says. "Most people know what a headphone plug

quite interesting. Many people think this LED looks like an album cover. One of my favorites has to be this electrolytic capacitor [above] — it has a neat spiral pattern inside!"

His process is pretty low-tech. Each component is simply sanded with normal sandpaper to reveal the internal structures. Sometimes, a piece may require the additional step of a hacksaw to take care of the bulk of the work, then sandpaper to finish. However, Schlaepfer sees a possible upgrade in the future in the form of a diamond saw and polisher.

What will he find next? Likely a publisher. "A lot of people on Twitter want to know if I'm going to do a book, so I am currently investigating that." —Caleb Kraft

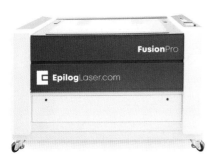

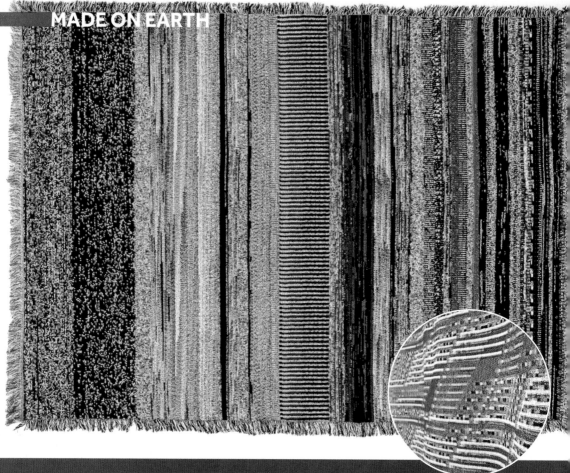

BLIPS AND BLANKETS INSTAGRAM.COM/GLITCHTEXTILES

Glitch Textiles

When Brooklyn, NY-based musician **Phillip Stearns** found circuit bending, he started experimenting with all kinds of art, branching out from audio into visual, then ultimately into textiles. That one tool, the ability to modify electronics hardware to create things from the resultant glitches, led to his endeavor called **Glitch Textiles**.

"The project started with an investigation of glitch images created by circuit bending digital cameras," Stearns says. "By deconstructing the hardware, it became apparent that the whole field of digital photography was more computational in nature, bearing little resemblance to non-numeric ('analog') photography."

Of course, harnessing glitches isn't always easy — their very nature means that the results are going to be somewhat unpredictable. However, their visual renditions are also often extraordinary, seeming like brief and fleeting moments captured in time. Vibrant colors and sharp contrast remind us of the digital underpinnings of the art, even as it drapes and flows over your living room sofa.

Initially, the glitches used may have all been created by Stearns' careful soldering and tweaking, but now there are many options, which come from some pretty interesting sources.

"There are still designs made with circuit bent cameras, but others use custom pixel sorting algorithms, or a custom data visualization I wrote, even exploiting rendering errors in Blender," Stearns says. "Most recently, I've been creating designs using malware that I've captured on servers I'm operating as honeypots, as well as visualizing recently discovered state-sponsored cyberweapons." —*Caleb Kraft*

AN INTERNATIONAL EMBRACE

ACADEMIC MAKERSPACES TAKE THE STAGE AT ISAM 2019

Written by Bill Weir

The "built" team shows Roya Mahboob, CEO of Afghan Citadel Software Company, their speaker kit, developed at Yale as a learning tool teaching the fundamentals of fabrication and electronics.

From practical shop talk to heady discussions on the future of making, the 4th International Symposium on Academic Makerspaces (ISAM) at Yale University provided a chance to trade notes, attend workshops, and listen to pioneers in the field. Co-hosted by Olin College and attended by 350 people from 156 universities from 14 countries, the three-day event featured workshops, makerspace tours, over 20 presenters, and six paper sessions.

In her presentation, littleBits founder Ayah Bdeir recalled the revelatory experience of discovering makerspaces, where she learned about tools, collaboration, and the value of failure. "The work everyone here is doing is so important," she said. "For me, it was critical — my life would have been very different if I didn't have that experience."

Closing the symposium was Afghan entrepreneur Roya Mahboob, who saw a computer for the first time at age 16. The "magic box" inspired her to become Afghanistan's first female CEO of a tech company. She led an all-girls robotics team from Afghanistan whose visa troubles in 2017 temporarily prevented their entry to the U.S. for the FIRST Global Robotics Competition. International attention from that incident led to work on a tech-based high school for girls and boys in Afghanistan. "Now it's time for Afghanistan's leaders to lay aside the misconceptions of the past and embrace the potential of Afghanistan's youths," Mahboob said in her appeal for the program's adoption. She is currently working with Yale on that effort.

Vincent Wilczynski, deputy dean of Yale Engineering and director of Yale's makerspace, the Center for Engineering Innovation & Design, points to what can happen when people come together with a common goal. "There was a question about what this community can do," he said. "It's actually anything this community wants to do."

BILL WEIR is the director of news and outreach at the Yale School of Engineering & Applied Science.

PROJECT COMMUNITY PROFILE: HACKADAY.IO

WRITTEN BY DAN MALONEY

DAN MALONEY is the community engineer and staff writer at Hackaday.

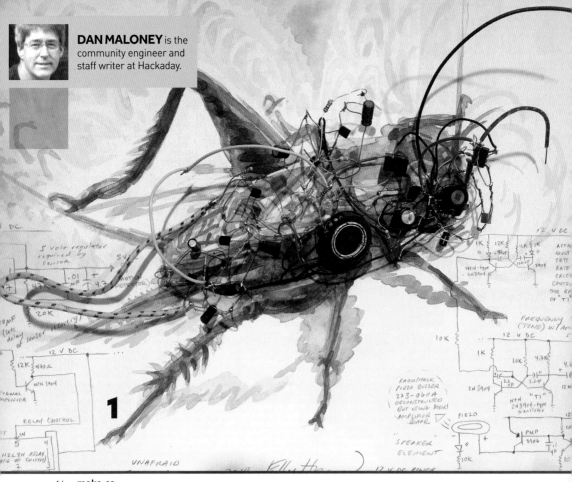

WE INVITED OUR FRIENDS AT **HACKADAY** TO TAKE US ON A TOUR OF THEIR PLATFORM, CONTESTS, AND ANNUAL GATHERING, AND TO SHARE SOME OF THEIR FAVORITE PROJECTS POSTED TO THEIR SITE.

Every day since launching more than 15 years ago, we at Hackaday have been writing about amazing feats of hardware sorcery found throughout the broad community. It became clear that readers needed a virtual place to call home, a place to document their own projects and connect with each other. So we launched the project platform Hackaday.io in 2014, and it has become the place to be if you're into hardware hacking. The site has grown into a vibrant community of 400,000 hackers, makers, doers, and thinkers with the passion to create and share. Members can build a team, engage with the community to get help, and celebrate successes and share the pain of failure. Here are a few of the almost 30,000 projects that really show off what it is all about.

1. MUSIC, ART, AND ELECTRONICS

Project: **Nature's Music**
Creator: **Kelly Heaton**
URL: **hackaday.io/project/161443**

Some Hackaday.io projects straddle the border between art and technology. In Nature's Musicians, Kelly Heaton uses electronics to replicate the sounds of birds, insects, and other non-human musicians, and incorporates them into sculptures and paintings of the animals. She'll be presenting her work at the 2019 Hackaday Superconference in November.

Kelly Heaton

THE HACKADAY PRIZE AND THE SUPERCONFERENCE

Launched in 2014 along with Hackaday.io, the Hackaday Prize has been a clarion call to hackers everywhere to put their skills to the test against the best in the world. Although the Prize differs every year in terms of theme, milestones, and requirements, it's always announced in the spring and runs through November. And the awards are substantial: We offered a trip to space for the first Hackaday Prize in 2014, and have settled on more cash prizes and perks since then. This year's grand prize is $125,000 and a residency at the design lab of Supplyframe, our parent company. The best entry from five categories will also win $10,000 each, and five honorable mention entries will get $3,000 a piece.

The Prize challenges hackers to create a project on Hackaday.io, assemble a team, and document progress as the year goes on. Entries are judged by the quality of the build, the impact of the project, how it addresses the theme of the Prize, and the quality of documentation. Winners are announced at the Hackaday Superconference in Pasadena, the premiere event of the hardware hacking year. The Superconference is where our community goes to connect, learn, engage, and network, while they're not diving into workshops or kicking back at one of the many social events.

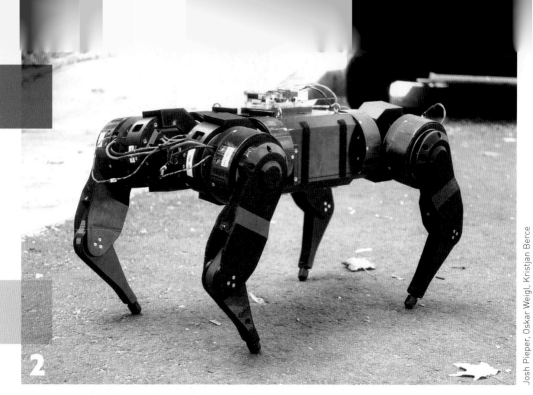

2. QUADRUPED ROBOTICS
Project: **mjbots quad A0**
Creator: **Josh Pieper**
URL: **hackaday.io/project/167845**

Josh Pieper is working on his own open-source robotic dog, similar to the one that Boston Dynamics is now offering for sale. But since it's open source and 3D printable, you won't need to have a Ferrari full of cash to afford it.

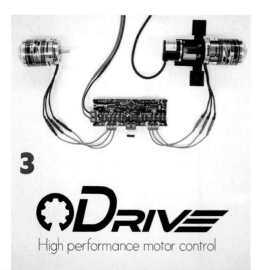

3. SERVOS FROM BRUSHLESS MOTORS
Project: **ODrive**
Creator: **Oskar Weigl**
URL: **hackaday.io/project/11583**

Robots like Pieper's need a lot of power at each joint, which is where the ODrive comes in. ODrive turns hobby-grade brushless motors into high-speed, powerful servos for CNC machines, 3D printers, and the occasional robotic dog.

Oskar Weigl's ODrive is just one of the many Hackaday.io projects that started as a Hackaday Prize entry and became a commercial product.

4. A HELPING ARM FOR THE ELDERLY
Project: **ExoArm**
Creator: **Kristjan Berce**
URL: **hackaday.io/project/20663**

Not every project is about building autonomous robots, of course. Embettering the world is a big motivator, too. This affordable exoskeleton arm by Kristjan Berce started as an attempt to help the elderly lift heavy objects; it too is now a 2019 Hackaday Prize entrant.

4

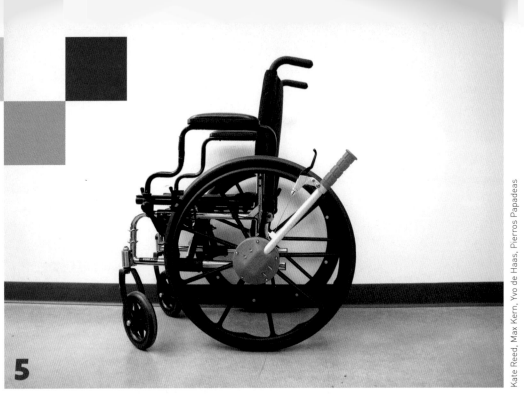

5

Kate Reed, Max Kern, Yvo de Haas, Pierros Papadeas

5. MAKING WHEELCHAIRS BETTER ON A BUDGET
Project: Hand Drive
Creator: Kate Reed
URL: hackaday.io/project/7221

With so many people relying on wheelchairs to get around in the world, Kate Reed thought there must be a better way to power them. So she built one: the Hand Drive. It's a wheel-mounted lever drive that's entirely 3D printed and costs only $40 to make, making it incredibly attractive compared to commercial versions.

6. TINY FPV BOTS FOR FUN AND LEARNING
Project: ZeroBot
Creator: Max Kern
URL: hackaday.io/project/25092

Speaking of the Hackaday Prize, here's an entry from the 2017 edition of the annual contest: ZeroBot. Max Kern's tiny, simple robot can be almost entirely 3D printed, and it makes a great weekend project to get kids interested in robotics.

A newer version, powered by an ESP32 and with low-latency FPV video, was also a submission in the 2019 Hackaday Prize. We love seeing designs refined and projects improved over the years.

7. A DIFFERENT KIND OF 3D PRINTING
Project: Oasis 3DP
Creator: Yvo de Haas
URL: hackaday.io/project/86954

Hackaday.io was big into 3D printing long before 3D printing was cool. And while you can now buy a 3D printer anywhere, our members are still pushing the envelope with new methods of additive manufacturing. The Oasis 3DP by Yvo de Haas is a binder-jetting printer that uses recycled inkjet cartridges to shoot a binding fluid onto a thin layer of powdered gypsum. The early prototype shows a lot of promise.

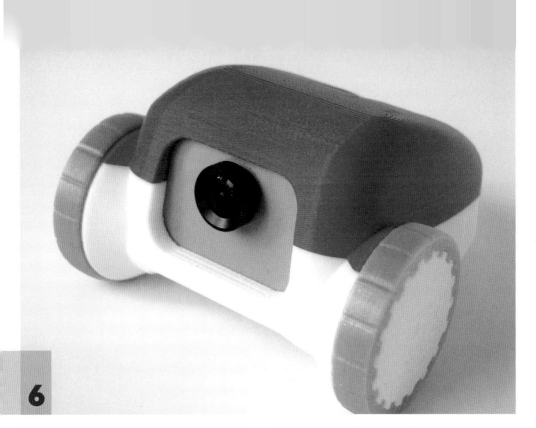

6

8. BRINGING SATELLITE TRACKING DOWN TO EARTH

Project: SatNOGS
Creator: Pierros Papadeas
URL: hackaday.io/project/1340

And finally, no Hackaday.io greatest hits list would be complete without a nod to SatNOGS, the Satellite Network of Ground Stations. SatNOGS won our first-ever Hackaday Prize in 2014, and while Pierros Papadeas and the SatNOGS team elected not to take the trip to space offered as a prize, they did manage to turn a lot of eyes — and antennas — skyward. SatNOGS has built out an extensive platform for tying together DIY satellite ground stations, which provides tracking data for hundreds of satellites. ⊘

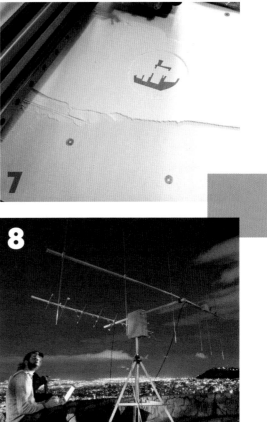

7

8

JIMMY DIRESTA

YOUTUBER PROFILE
WRITTEN BY CALEB KRAFT

WITH PENCILS, STENCILS, AND HIS TRUSTY BANDSAW, JIMMY INSPIRES MAKERS EVERYWHERE

DIRESTA is a name that has become synonymous with "maker videos." Some see it and think maybe it is a tool brand or a network name, but it is the last name of Jimmy DiResta, a maker based out of New York.

There's a good chance you've seen DiResta's videos. From his Discovery TV show years ago, to his own YouTube channel, to the countless others that he has contributed to (Make: for example), his work has spread far and wide. His telltale style is purely focused on the build. He doesn't typically narrate or talk you through his motivations, you simply see his hands working, feeling their way through each project to get to his desired goal. Time is a flexible thing in a DiResta video, with some sections going hyper fast, and sprinkles of slow motion to accentuate a point. It marries together into a mesmerizing effect that keeps you streaming video after video of his. Others have started replicating it but he still does it the best.

In discussions with Jimmy, you'll find that he is silly and playful in behavior, something you can easily notice in his projects. He doesn't have classical training to build what he does, nor does he typically have a plan. Rather, he just begins, takes each problem as they come, and tries to be creative in his solutions. Even his editing can show this with bits of humor sprinkled throughout.

With 1.6 million subscribers at the time of writing, DiResta's YouTube channel is not the absolute largest in the maker community, but when you see Jimmy at an event you understand that he is a true celebrity. Crowds gather, lines form, and yet Jimmy is always humble and gracious, willing to talk and swap stories.

We caught up with him to learn the ins and outs of how he produces his ever-popular clips.

MAKE: What got you into making videos?
DIRESTA: My channel started as a reaction to my show (*Dirty Money*, 2011) getting cancelled on the Discovery Channel. I was trying to show off and

let the TV people see what they're missing — they never really seem to pay attention but the fans did and that's what really matters in the long run.

What makes a build a "DiResta" build?
When I began making videos I knew that I needed to make a lot of them so I avoided using a voiceover or trying to string together a series of explanations. I knew that it would slow me down in the edit. I strictly stuck to telling a story with visuals and I believe that's what set me apart early on and established what a "DiResta build" is. Also I was never overly concerned with the audio. The sound of the tools is what I used. No music or complicated copyright decisions to make.

What is the big "ooooooh aaaaaah" shot for your videos? The one that viewers wait for?
I guess it's when I go from superfast to super slow; many people always expect an accident but most often I'm just trying to highlight something that I want people to recognize or a technique that I need to slow down so people understand.

CALEB KRAFT
is senior editor for Make: Community and now, almost exclusively, uses YouTube for research, education, and inspiration.

Mark Adams

1

2

STRICTLY STUCK TO TELLING A STORY WITH VISUALS

3

4

5

6

How much do you plan your videos?

You might say I storyboard it all inside my head while I'm making any video or product. I keep asking myself what would I want to see next and this helps me lay out the video in real time while I'm working on the project.

What influences your project selections?

I choose projects based on what I'm feeling at the moment. I don't typically plan a video just because I know it will do well and most often plan a video because there is some personal challenge in it that I need to overcome or get good at or learn something about.

Give us a couple "Here's a cool trick" tidbits.

A thing learned while editing hundreds of videos is shoot very tight. I see far too many beginners shoot every single part of the process and find it impossible to make a 10 minute video out of 16 hours of footage. Only shoot when material changes from one shape to another.

Another funny edit trick I use often is when something has a left and a right side (symmetrical stuff like tables and canoes) I often just shoot one side of it and then flip the image in the edit, it makes it seem like I went from the left side of the project then I did the same thing to the right side of the project. Many times words and label appear as a mirror image and no one seems to notice!

How do you handle negative comments?

Out of respect to my advertisers (I call them clients) I try to keep my responses upbeat and funny, and never call anybody nasty words. But if somebody really comes at me I simply delete the comments and block them. I never let a thread go too far before I just simply delete it.

What would you say to someone wanting to start making videos about making stuff?

You just have to start. I see too many people waiting for the giant moment of inspiration that's going to set them apart. This is something that you need to do over and over and over again to find what is absolutely unique to you in your process. Don't just wait for it to happen because it never will unless you're actually working at it. ⊘

> [TV PEOPLE] NEVER REALLY SEEM TO PAY ATTENTION BUT THE FANS DID AND THAT'S WHAT REALLY MATTERS IN THE LONG RUN

1. Jimmy received a hand-written request from a young viewer to make a batarang. It even included blueprints! This was his result.

2. Jimmy's laser-cut, hand-sharpened steel knives.

3. A group gathers at DiResta's upstate NY farm for classes in various maker skills.

4. Steel and oak end tables. DiResta's contribution to a fundraiser auction for Boys Town.

5. Jimmy working the bandsaw at the Louisville Maker Faire.

6. A leather tool bag that DiResta made for the Weaver Leathercraft YouTube channel.

Calling All Climate Makers:
Electrify Everything!

Written by Saul Griffith

THE CHILDREN ARE RIGHT: THIS IS OUR LAST CHANCE FOR A LIVABLE PLANET. WE NEED AN EPIC EFFORT TO ELECTRIFY VEHICLES AND ENTIRE HOMES. MAKERS CAN MAKE IT HAPPEN

Stand with Your Children

This is the first of 4 articles on makers and climate change. This one's the big picture — why it's urgent and why it's about technology substitution, not efficiency.

No one knows how to get to zero carbon; it will take thousands of innovations. In future issues we'll tackle the 3 big areas where Makers can make the change: our vehicles, our homes, and our stuff. We'll dive deep on electric vehicles, how to electrify entire homes and why that's even more important, and decarbonizing all our stuff (and keeping plastics out of the ocean while we're at it).

Adobe Stock - sergeyvasutin

The future *could* be awesome. Solving climate change and improving everyone's lives in the bargain is possible. But so far it has proven politically impossible. Our climate policies and Green New Deals are frankly not enough to hit the 2030 targets that science tells us are necessary. We will likely lose all the coral reefs, suffer intolerable ocean acidification, and set in place carbon feedbacks such as methane emissions from melting tundra, if we allow the planet to warm any more.

America can utilize its fabulous natural resources to provide abundant zero-carbon energy to all its citizens. Energy will cost less than ever before, and new jobs will be created in every zip code. We'll enjoy cleaner air and water, rejuvenated agriculture, and better food. Once on the path to zero emissions, we'll thrive by exporting technology and know-how to the world.

That could be the future we live in, and it is worth fighting for. This article is about envisioning the project to decarbonize America (and the world), and the role of makers in achieving it.

It can be done. It is urgent. It is audacious. We need to stand with the children and get it done, because right now they're the only adults in the room standing up and telling it like it is. It is a climate emergency. Extinction Rebellion and the Youth Climate Strikers aren't extremists. They're the only groups showing appropriate urgency.

We pissed away three decades fighting with ourselves and allowing fossil fuel companies and climate deniers to distract us. We don't

DR. SAUL GRIFFITH is founder and principal scientist at Otherlab, an independent R&D lab, where he focuses on engineering solutions for a clean energy, net-zero carbon economy. Occasionally making some pretty cool robots too. Saul got his PhD from MIT, and is a founder or co-founder of makanipower.com, sunfolding.com, voluteinc.com, treau.cool, departmentof.energy, materialcomforts.com, howtoons.com, and more. Saul was named a MacArthur Fellow in 2007.

actually have "10 years" as implied by the headlines. We've already deployed infrastructure that will burn enough carbon to take us well past 1.5°C and maybe past 2°C. That means we need to replace every vehicle, power plant, furnace, and stovetop, with a zero carbon option, the next time it's being replaced. We need 100% adoption and perfect execution.

Game on, makers, it's your time to shine. Not only do you need to invent all the widgets, but you'll need to start advocating for the right solutions, and being ambassadors for the better world we could live in.

Why Efficiency Can't Cut It

The last time we had a major energy crisis, in the 1970s, the problem was oil imports that put the U.S. economy at the mercy of foreign cartels. At the same time, air and water pollution caused by our energy production also came to the fore, thanks to a burgeoning environmental movement inspired by books like Rachel Carson's *Silent Spring*. Both political sides came together, and the Environmental

U.S. Energy Flow, 1976: Primary Consumption 72.1 Quads

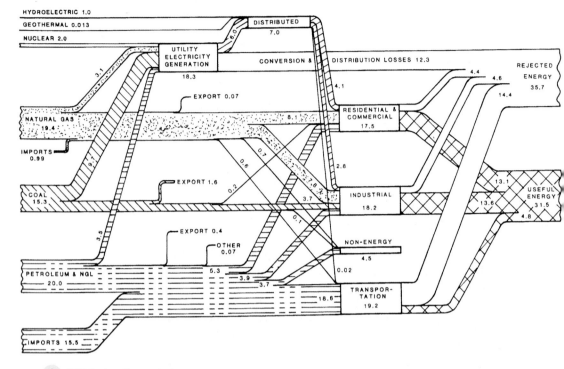

A 1976 Sankey diagram by Lawrence Livermore National Laboratory. U.S. energy use was about 72 quads.

Estimated U.S. Energy Consumption, 2018: 101.2 Quads

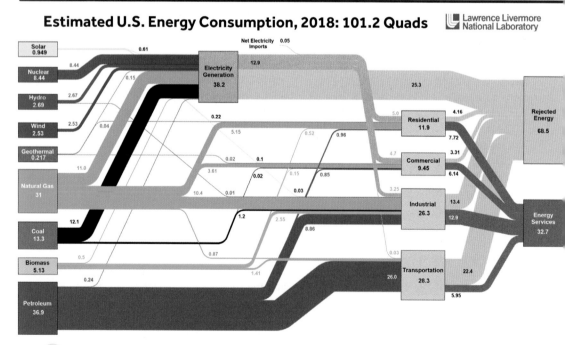

B 2018 Sankey diagram by Lawrence Livermore National Lab. U.S. energy use is now 101 quads, most of it waste.

Protection Agency (EPA), Department of Energy (DOE), and Energy Information Administration (EIA) were brought into existence by Presidents Nixon, Ford, and Carter. At that moment in history, it was conceivable to solve these problems by efficiency measures (to reduce foreign oil dependence) and by regulations (to control emissions).

Bipartisan technocracy (letting engineers and scientists show the way) was key to focusing policy makers and the public on the problem and the solutions. The EIA created the first Sankey diagrams analyzing the total energy flow in the United States (Figure **A**). The problem identified in 1976 was the "IMPORTS, 15.5" component at the bottom left. Out of the 72 quads (quadrillion BTUs) of energy we used, 15.5 quads were imported. We developed the CAFE auto fuel standards, appliance standards, and an efficiency mindset to solve this problem. In concept, efficiency could still save us.

We have a new energy crisis that's different, but we're trying to solve it with the same old ideas. Indeed, the 2018 Sankey diagram (Figure **B**) looks basically the same as 40 years earlier. The majority of our energy sources still produce carbon dioxide as a by-product, an emission that's causing the climate to warm with deleterious effects. We're trying to fight this CO_2 problem with efficiency and regulation. Unfortunately that plays directly into the hands of those who would continue to emit, because efficiency sounds to the voter like "less" and regulation sounds like "big government."

You can't "efficiency" your way to zero emissions, and voters know it. What we must do instead is *substitute* better technology. We don't want less, we want an improving quality of life. We want more, faster, cleaner, healthier, better. The correct approach is to look at every single way we use energy, and find a better way to perform those services that is zero net carbon.

Electrify Everything

The fastest way to zero emissions is through electrification. Makers know this in their guts — electric machines are so much more efficient than their fossil fuel-burning counterparts that we'll end up with more, not less, without even thinking about "efficiency," but just by committing to electrification and infrastructure renewal.

» When you drive a car using gas, more than two-thirds of the energy is lost in burning the fuel. Electric cars powered by renewables or nuclear use less than one-third of the energy per mile of their dinosaur-powered counterparts.

» Electric heat pumps need just one-third as much energy to heat a home as a natural gas furnace.

» Using fossil fuels to make electricity means almost 60% of the energy input to the grid is wasted as heat.

When we electrify transportation and building heating and cooling, when we power the grid on renewables (and nuclear), **we will need *less than half* of the energy input to the economy that we use today** (Figure **C**, next page). That's the only efficiency measure you need — electrification.

Our cars could be just as big, only electric. Our homes just as large, only electric. Our economy just as big, only electrified. In fact, the electrification of everything will allow us so much extra productivity in the economy that we'll probably be able to have more of some things. The American Dream could be better than it ever was.

Here is a defensible and believable pathway for politicians of both sides, a vision for America that's bigger, better, faster, cooler, more powerful, and that *bests* the climate commitments of Paris. (Why merely meet Paris goals when we can *crush* them?)

Better Tech in Every Home

It won't be easy. We need to electrify 200 million vehicles, replace 80 million furnaces and 90 million water heaters, and put solar on 100 million roofs. It is possible, and can be done at the natural rate of replacement of these items (cars on average are 13 years old, furnaces 20, asphalt shingle roofs 15). But we need to start immediately, and every one of them needs to be electric.

Not only the grid needs to be upgraded, but the infrastructure of our households as well. This brings the climate change conversation right to the kitchen table, and it's a positive conversation too. Your house will be warmer and more comfortable when we shift to heat pumps and hydronic heating. Your cars will be faster and safer when they're electric. Household air quality will improve, as will our health — natural gas in homes is a respiratory problem for children and pets. At the scale of the economy, costs will come down and every family will save thousands of dollars a year.

U.S. Energy Consumption, If Electrified: 40 Quads

C New scenario Sankey diagram: Same U.S. economy, if we electrify everything, uses 40 quads, *less than half* of today's energy usage; major savings highlighted at top right. Drill down into the diagram details at departmentof.energy.

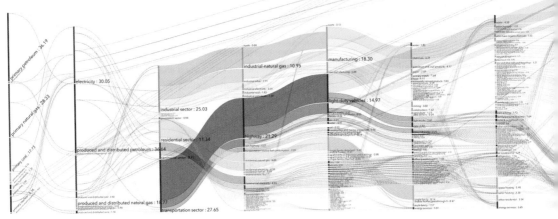

New Jobs in Every Zip Code

Most Green New Deals are slight in detail, and all need to be more ambitious. And they all miss the opportunity to connect climate change solutions to our households, and to a grand vision of American renewal. This is again because we mostly think about the supply side of energy: where it comes from. We need to expand that 1970s vision to include the demand side: the end uses, how we actually use energy in our homes and businesses.

Doing this vastly expands the number of jobs that will be generated, and those jobs will be in every zip code. There's an awful lot of work in putting solar on roofs, replacing furnaces with heat pumps, and rewiring the home to charge the electric vehicles in the driveway. We need to connect all those homes with micro grids, build vehicle charging infrastructure everywhere, and connect the entire continent to move energy long distances to meet 24/7 demand. There are even more jobs in decarbonizing industry and developing climate-friendly agriculture.

We recognize that most fossil fuels (85%–90%) are mined and extracted in large, resource-rich states that tilt "red." But it's these states where the majority of utility-scale wind and solar will be built, because they have the land area. Some industries will lose, but those are the industries who have been lying to you and your children for decades. Entirely new industries will spring up.

If America acts first, we'll be one of the major economies that export this clean infrastructure to the rest of the world. The U.S. needs to lead, and just as surely Japan, Germany, South Korea, and China will follow. The bigger the commitment we make to leading the world, the larger the share of the export pie we'll hold as we produce the infrastructure of the future for the world.

U.S. Finance to Unleash the Future

There is one thing government can do that no one else can: provide the financing backdrop for this revolution. We've done it before. As part of the New Deal, Congress created the Federal Housing Administration to enable the government to release capital to local banks to finance housing loans. The act was amended in 1938 to create Fannie Mae. This influx of capital enabled the suburbanization and housing boom of post-WWII America by enabling far greater access to low-interest loans. Today, slashing energy bills would increase a family's cash flow and capacity to finance their home ownership. At an 8% interest rate financing solar cells, half the money goes to financing. At 3% it's negligible. The government could find enormous leverage in the private markets with a low-interest infrastructure "climate loan guarantee" instrument. Every voter would benefit from the lower energy bills and renewed infrastructure that results.

Super Sankey 2018: Saul Griffith & Keith Pasko | www.otherlab.com, departmentof.energy

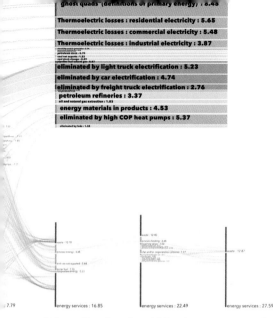

Solar cells, electric vehicles, heat pumps, and batteries mean that the balance of infrastructure is shifting to the consumer. Providing government insurance or home loan-quality financing will enable consumers to purchase this infrastructure and lower their energy bills — infrastructure that will mean high-quality jobs for decades. And a more distributed grid built on a profusion of interconnected micro grids will be much more resilient than today's electric grid, just as internet infrastructure is strong because of its distribution.

The Decarbonization Revolution

Some people have likened this decarbonization project to the Apollo moonshot, others to Democracy's Arsenal, the manufacturing buildup for World War II. Some say it's a Manhattan Project in scientific scope. Some think it's a new New Deal.

To have a chance of creating the future our children want and need, it is all four of these at once. The manufacturing buildup of electric vehicle production, solar plants, and wind energy is akin to Democracy's Arsenal. Figuring out how to sequester carbon cost-effectively is a Manhattan Project (and developing clean commercial fusion energy is another). The new New Deal is the grid build-out required for such an enormous shift to an electrified economy — the transmission lines and the charging infrastructure. Want a moonshot? Figure out how to sequester carbon in

better agricultural practices while feeding more people better food. Or how to decarbonize industry and use synthetic biology to replace plastics.

Mid-20th-century America was built on an audacious combination of science projects, visionary infrastructure, innovative manufacturing, and novel financing, all supported by and in partnership with government. **It's why the world looks to America to lead the decarbonization revolution — we're the only country with a history of achieving projects this ambitious.**

Makers Will Make It Happen

Waiting 4 more years to implement such a plan is likely the difference between a 1.5°C and 2°C warming of the world. That could be the difference between some coral reefs and glaciers, and none at all. A plan this audacious and American could've come from either political party in the mid-20th century. Perhaps it could come from either party in 2020 but I'm scared it will come from neither.

The best way to demonstrate the future by is building it. This is where makers can shine. All you early adopters, hackers, inventors, designers, and activists — now is the time to design, build, and implement the solutions for a decarbonized world around your own house, connected to your community, as shining examples for your county and state, that fit into the national picture, to create an international solution. There's a lot to learn from each other: what works for different regions (HVAC solutions differ wildly by climate region) and population densities (carbonless transport in cities looks very different from rural areas).

It's in our hands. Climate change is now with us, and with us forever. Honestly the challenge is to design and popularize a world and a way of living in it that doesn't descend into climate chaos or negative feedback loops, doesn't destroy more mammal life, marine life, or critical oxygenating rain forests, and provides a happy environment for 8 to 10 billion people to prosper.

For the first time, I think we can squint and see a solution where that decarbonized world is possible, without requiring us to invent something completely magical, like fusion. The biggest question is whether we can summon the political will, and the concerted effort, to make it happen. ⊘

WORLD'S FIRST
CHEESEBURGER

A **Bun conveyor:** Pneumatic piston, laser sensors, and control algorithm push whole brioche buns single file into the ...

B **Bun slicer:** Sensors detect bun, then servos pull wood block down to push bun through oscillating blade. Whole buns stay fresher than pre-sliced, don't need preservatives.

C **Vertical toaster:** Butters the bun heel and crown, then camshaft-operated paddles press them onto the two-sided vertical griddle, and scoot them along to the ...

D **Case handler:** Vacuum-plucks a paper clamshell box off the stack and places it on butterfly belt, where it catches the toasted bun.

E **Butterfly belt:** Key to moving the burger down the line and reading its position at all times. Unlike ordinary conveyors, any portion of this belt can move independently.

F **Saucer:** First stop inside the refrigerated assembly line. 15 different condiments are squirted by a Cartesian X-Y sauce head similar to a plotter.

G **Veggie slicers:** Pickles, tomato, and onion are sliced fresh onto bun heel.

H **Lettuce portioner:** Pre-shredded butter lettuce is the only part of this burger not cut to order seconds before you eat.

I **Cheese shredder:** Grates two types of cheese directly onto bun crown. Load cells in receiving cup measure portion to nearest gram.

BURGER BOTICS
Want algorithms with that? The Creator robot is made of a dozen-plus modules, 20 Linux computers, and 350 sensors, each playing its part in an impressive culinary choreography.

KEITH HAMMOND is Senior Editor of *Make:* and does love the taste of a good burger.

ROBOT

HOW TWO CALIFORNIA KIDS OVERCAME DOUBTERS (AND INTERSTATE 5) TO AUTOMATE THE FRESHEST BURGER EVER SERVED

Written by Keith Hammond

Aubrie Pick

J **Cheese melter:** Like an intense heat gun. Initially had noise and vibration issues, solved with mounting solutions from automotive industry.

K **Seasoner:** Pneumatic puffer blows spices onto burger or bun, depending on recipe. Custom acrylic hoppers with 12 seasonings are rotated into position by proprietary gearing system.

L **Grinder:** Custom built, refrigerated unit grinds marinated brisket and chuck fresh to order — no oxidation — and delivers it to the ...

M **Patty former:** Secret technology, inspired by chef Heston Blumenthal's technique, aligns the cut direction from the grinder vertically with your bite direction, for the tenderest bite. Detects portion amount, then briefly smashes patty for optimal sear, and releases to keep it juicy.

N **Induction grill:** Double electric contact griddles cook the patty top and bottom simultaneously, 40% faster than single grills, using 11 thermosensors and AI machine learning algorithms. No two burgers are cooked exactly alike.

O **Triple air filtration:** No grease, smoke, or smell — only warm air comes out.

P **Robo spatula:** Scoops patty from grill onto bun. A visible-light proximity sensor alerts the staff that your Creator burger is complete.

Aubrie Pick, Creator, Omeed Manocheri, Stanford Product Realization Lab

It's San Francisco, so naturally there's some controversy. Burgers are the number one segment of restaurant food, ripe for disruption by robotics, and the VCs are pouncing. Robots might take away fast-food jobs, or create better ones, so the labor activists are activating.

But today I just want to understand how they made it. How two young engineers could invent a complicated culinary robot that makes cheeseburgers from scratch. Really good ones.

I'm on my second visit to Creator, the world's only robotic hamburger restaurant. Since opening doors in September 2018, it has built a rep for quality and innovation, thanks in part to its fun-to-watch robot and its tourist-trod location next to San Francisco's Moscone Convention Center.

But the real secret sauce is the small team of makers who overcame technical setbacks and ridicule to create a single, integrated robot that makes the freshest burger ever served. And for just 6 bucks! Which even a tourist knows is unheard of in San Francisco.

Everything in a Creator burger is prepared to order in a way that's never been done — beef freshly ground 20 seconds before it is precision grilled by artificial intelligence; whole buns automatically split, buttered, and toasted; veggies precision sliced and cheese shredded the instant before they meet the robo-toasted bun. And yes, it's delicious, thanks to quality ingredients and gourmet techniques and seasonings.

Today the burger bot team includes alums from Tesla and Google, Berkeley, Stanford, and Caltech, Chez Panisse and Fat Duck. But it started with two California kids barely out of college: a self-taught roboticist with a vision for the future of fast food, and a trained mechanical engineer who came to trust in that vision. I want to hear how they created the culinary equivalent of landing a moon rocket on a robot barge.

BOOTSTRAPPING ON I-5

Alex Vardakostas grew up flipping burgers in his family's restaurant in the surfy Southern California town of Dana Point, and earned his physics degree from UC Santa Barbara. His dad, a Greek immigrant, saved enough to buy his own burger joints in the 1970s. His mom, also an immigrant, lost her engineering career to the Iranian Revolution and became a burger chef.

During college Alex became obsessed with the idea of completely automating the cheeseburger. He believed a robot-made burger could be faster and more sanitary, but above all, fresher and tastier, because a robot could perform modernist gourmet techniques while also freeing up money for better ingredients.

After graduation Alex began building his robot, teaching himself mechanical engineering as he went. Without prototyping tools of his own, he became a regular at TechShop, the late lamented chain of open-access DIY fabrication workshops — but this required brutal 800-mile round trips on Interstate 5.

"I built the first robot in my parents' garage in south Orange County," Alex says. "There was no TechShop in L.A. then, so I would drive all the way from Dana Point to the Menlo Park TechShop to do parts fabrication, crash on a friend's couch for weeks, then drive it back to Orange County and see if it fits! It was pretty lonely driving the 5, listening to Stanford ETL lectures on audiobooks. I grabbed a bunch of textbooks — Shigley's *Mechanical Engineering Design*, *Practical Electronics for Inventors*, *Machinery's Handbook* — and *Make: Magazine* was inspiring."

People just didn't get it, Alex says. "I spent two years in the garage working alone, trying to push it along, to solve these problems of how robots need to meet food, which is squishy and variable. Everyone's telling me, 'You're crazy. Robots can't do this, there's too much dexterity required.' I found it more socially acceptable to say I was unemployed, rather than working on a cheeseburger robot in my garage."

Learning robotics from scratch was a steep climb for Alex. "I had a physics background but that's not engineering. I didn't know stress from strain, or what a microcontroller was. The first day I learned what an endmill was the first day I used the CNC — that's where I met Steve, at TechShop, cranking on the mill!"

AN ENGINEER APPEARS

Steven Frehn was raised in the no-nonsense city of Palmdale, California, a military and aerospace center in the West Mojave Desert. In high school

Alex Vardakostas with his parents, Angelo and Maheen, at Burger Stop in San Clemente, California, mid-1980s.

First-gen cheeseburger robot built by Vardakostas with help from Steve Frehn, in the family garage in Dana Point.

Creator cofounders Alex (top) and Steve (bottom) today. Steve (right, at Stanford, 2007) brought fabrication chops to the project after meeting Alex at TechShop in 2010.

Burger engineers, 2012: (left to right) Jack McDonald, Alex Vardakostas, Steven Frehn, and Ari Atkins.

The second-gen burger bot, before the grinder, patty former, or grill were developed.

Old-school conveyor belt in 2012, the type responsible for at least one mortifying *I Love Lucy* incident.

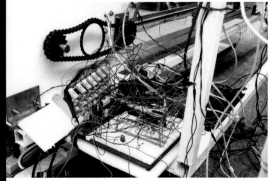

2012 prototype, stuffed with 8-bit PIC microcontrollers and at least one giant breadboard. Today it's all Linux.

he interned for NASA at Edwards Air Force Base and placed third in a national science contest to win a scholarship to Stanford, where he earned a degree in mechanical engineering. In 2010, Steve was making parts for his day job designing rooftop solar panels, while volunteering at Menlo Park TechShop in exchange for membership. A cheeseburger robot was the last thing he expected to work on.

But Alex outlined the burger bot dream with his usual enthusiasm and culinary expertise, and soon Steve was milling parts at TechShop and FedExing them down to SoCal, where Alex tested them and then sent requests back for modifications. They started hanging out to do design reviews and share technical challenges.

Steve brought the hands-on fabrication chops and maker DNA the project had been waiting for.

"Two years later I had a GUI on my computer, plugged a USB cable into it, and a burger would appear in my garage!" recalls Alex. "My best friend came over and it was weirdly quiet, like we knew that this had actually happened — it was possible."

VENTURE TIME

By 2012, Creator had a little incubator money. Alex moved the startup to San Francisco as CEO, and Steve took the leap to join full time as COO. Joined by two more engineering grads out of Berkeley and Utah, they prototyped their second robot at TechShop's three-story fabrication palace in San Francisco. Seed and venture money from Root Ventures, Khosla, and Google followed, and in 2014 the duo assembled a team of engineers from Tesla, Google, Boston Dynamics, and Walt Disney Imagineering.

Today the fourth-gen robot team is led by systems engineer Noe Esparza, a Stanford PhD roboticist (and long-ago co-intern at NASA with Frehn); hardware engineer Michael Balsamo; software engineer Jeffrey Jensen; and R&D engineer James Brinkman, ex-Navy and proud to be Creator's "first Craigslist hire!" I quizzed the team about the robot's development, how it all works, and memorable roadblocks or failures.

OVERCLOCKED DORM FRIDGE

"I like the culture here, failure is seen as an opportunity," Michael says. "I came here from

I LOVE LUCY

In 2015 team had 24 hours to move their latest prototype robot and set it up for an investor demonstration at Cavallo Point, a swanky resort across the Golden Gate Bridge from San Francisco.

"This was when we still had a regular conveyor belt," says Noe, "before the butterfly belt. It was straight out of *I Love Lucy* — suddenly our conveyor turns on full speed — and backwards! We're running around trying to turn it off, and burgers are like, *thunk, thunk, thunk* against the back wall!"

"Everyone looked at me!" yelps Jeff. "They thought it was a software issue, but Noe found out it was a bad wire crimp that got shorted when we moved the machine."

Alex chuckles. "The bacon aioli was really good on that one."

MASSIVELY EMBEDDED COMPUTING

Presumably this giant burger beast is stuffed with microcontrollers? Not anymore.

"Initially we used a lot of Microchip PIC micro-controllers, but that was making it more difficult to be modular, to iterate, to debug," says Alex.

"Now it's got 20 computers in it," Jeff says. "We're using mostly BeagleBone Blacks, and it's all Linux, off-the-shelf Ubuntu. Unlike most embedded design for mass manufacture, we didn't have the same cost and space constraints, so it's faster to use Linux throughout, and much easier to work with the mechanical engineers; we don't have to translate down to 8-bit microcontrollers."

Creator relies heavily on open source software for robot communications (Apache Thrift and *libserial, curlpp,* and *Boost.Signals2* libraries) and for development (Docker, Eclipse CDT, CMake, gtest, Jenkins, React Native, and Atom). And like a lot of makers on a budget, they still design circuits in EagleCAD (and these days, Altium) and order their purple prototype boards from OSH Park.

GRILLED BY AI

There are a lot of proprietary bits the team couldn't show me. In particular, the burger grilling module is still a black box, but the results are impressive. (I'm a flame griller at home but Creator's tasty burger makes a strong case for griddling instead.)

"It's loaded with temperature sensors and really

Tesla and it was the same way there. Figuring out refrigeration on this machine was hard, with the huge transparent assembly area that's got these openings on the ends. We built a prototype, but how do you measure airflow? We thought we'd do it like the automakers do, so we got a smoke candle firework and tested it on the roof ... "

"And it makes this huge cloud of smoke and San Francisco Fire Department shows up with a ladder truck!" Jeff laughs.

Steve looks anxious just remembering it. "We explained, 'We're a food technology company, we're testing a prototype, we apologize ...'"

"They were pissed," recalls Jeff with a grin.

"Then we were testing the grinder performance, and they don't perform so well when they're not refrigerated," Michael continues. "So we had to go to Lowes —"

"Dude, you took ours!" interrupts Alex. "We had a dorm-room little fridge and Michael took that ..."

"Oh yeah ..."

"And gutted it."

"We gutted it and put it on the grinder, added a shelf fan for active cooling ..."

"To overclock the dorm fridge."

"And installed aluminized fins on the evaporators to really sink the heat," Michael continues. "We forgot to turn it off once, came back and it was a giant block of ice!"

sophisticated algorithms to get the right doneness — no two burgers are cooked exactly the same way," Alex says. "There's some machine learning involved." Moisture sensors too? Probably but they would neither confirm or deny.

They did share the saga of inventing it. The first prototypes were built with gas burners. "To get enough gas ribbons and starters, we bought three gas grills at Lowes, and stuffed all the burners inside our grill unit," says Steve. "It got so hot it would singe your hair just looking in it, and wires from 3 feet away would start to melt."

Electric griddling was much easier to control, so they started over from scratch. "The good news is, griddling gives us a much better burger," says Alex. "You get better heat contact and a better crust, that Maillard reaction above 310°F. It's way juicier now. And no flare-ups."

It's also far more efficient. "We do electrical heating of griddle surfaces only when there's a burger on it. A regular restaurant's griddle is on all the time, the whole surface is hot, there's a lot of waste. We may be running the world's only all-electric burger joint," Alex says proudly. Inside are 11 thermosensors watching ambient and griddle temperatures, constantly adjusted based on machine learning AI algorithms.

The air filtration is impressive, too. The air is weirdly, impossibly fresh inside this burger joint. Yes, there's the expected grease baffle above the grill — it's required by building code. "But we built two extra air filters into the robot," Alex says, "one for smoke, and one for smell. The only thing coming out of the robot is warm air."

BUNS VS. STEEL

Like many parts of the robot, the bun system is on its fourth major revision. "The first robot had the buns stacked vertically in two hoppers, heels and crowns, and an arm would whack the bottom one into the toaster. In general there were a lot of issues with squishing," says Alex.

"So we rejected that and tried a single horizontal feed for the bun slicer," says Noe. "We found that a Gatorade bottle was the exact diameter of a bun, so we filled it with different amounts of water

> **"OUR FIRST SLICER HAD THE BUNS STACKED VERTICALLY, AND AN ARM WOULD WHACK THE BOTTOM ONE INTO THE TOASTER. THERE WERE A LOT OF ISSUES WITH SQUISHING."**

to test the weight." The horizontal feed is fun to watch. More importantly the buns aren't smashing each other and they stay fresh until sliced to order.

Noe explains the solution. "It's pneumatic: air is pushing the wood block piston, with lasers to read the position of the bun at the end, providing continuous feedback to the air system."

Alex elaborates, "The normal way is a conveyor belt, but that has lots of moving parts, and we wanted something easier to reload. We tried sucking the buns with a vacuum, we tried blowing them with air, but that would dry them out. We were using a Shop-Vac and manifolds we made from MDF at TechShop."

"Then we came up with the idea for a pneumatic pusher piston," says Noe. "That worked as far as the mechanics, but then there's a lot of algorithmic stuff you've got to do, the sensors and reactions."

"There are industrial slicers that slice a lot of buns all at once," Alex says, "but one-at-a-time-slicing, to order? It's tricky. Nobody had done it before. There was no place to start."

The bun slicing unit uses infrared and laser sensors to detect when each bun has arrived, then actuates the blades as servos move another wood block piston to push the bun through the blades.

ROBO TOASTER

The two-sided vertical toaster conveys the bun heel and crown through a buttering station — where secret technology applies liquid butter evenly — and three griddling stations. Alex chimes in with the food nerd science: "A little layer of butter really helps thermal conductivity, caramelization, and crispiness."

The buns are moved along and pressed onto the griddle by a series of 16 paddles, all actuated and timed by a camshaft that wouldn't look out of place in your car transmission except for the beautiful wood and aluminum design. Noe roughed it out in CAD and then iterated by hand, cutting linear cam profiles and feeding them through the paddles to observe the resulting motion and model a final sequence that toasts the bun optimally and also looks good to watch.

David Kover, Aubrie Pick, Creator

Like most of the robot, the toasting system is the product of trial and error. "First we tried an infrared toaster off the shelf, then we built our own infrared toaster," Noe says. "Then finally we built our own contact toaster."

Four "handoff paddles" at the bottom of the toaster drop the crown and heel, butter side up, into their respective halves of the open clamshell box, which Creator calls a "case." On our first visit this drop sequence was highly accurate, but not without the occasional inverted bun. The team is continually fine-tuning the paddle motions using motor position feedback data from the servos.

MOVE IT STOP IT
SAUCE IT SLICE IT

The saucer today is a far cry from the three upside-down condiment bottles in the early prototypes. It dispenses 15 different sauces including of course homemade catsup, mustard, and Pacific Sauce (a California twist on your usual Thousand Island-y burger sauce) but also charred onion jam, sunflower seed tahini, shiitake mushroom sauce, and oyster aioli. Positive displacement pumps, monitoring flow to the milliliter, are a hybrid of off-the-shelf systems and Creator's own technology; challenges included dealing with different viscosities and particulates. Sauces are distributed in a 2D pattern on the bun by an X-Y "sauce head" that moves just like a plotter or 3D printer head.

At the next station, washed, trimmed whole veggies in three hoppers (onion, tomato, and pickles) are gravity fed to blades that sense their own position, then sliced to precise thickness. But veggie slicing was a thorny problem for years.

"[Root Ventures founder] Avidan Ross said the proper thickness of a sliced onion for a burger is so thin you should be able to read through it," Alex recalls. "Once we got it dialed, we sent him a picture of a slice over the top of a message: 'Is this thin enough?'"

Slicing a tomato may seem simple, Alex says, "but we've working on the slicers longer than probably anything else. A lot of IP goes into getting that last slice! I can't even do it right at home."

So how much IP is in this robot? "We're probably at two dozen patents at this point, and a lot of them are really broad 'omnibus' patents," Alex says with a gleam in his eye. "No one's ever done this." ⊘

Sauce dispensers: 2012 (top left) and today (top right).

Veggie slicers: Rev. 1 (above) and today (below).

BUG GRUB

**Written and photographed
by Rima Khalek**

INSECT MEAL IS A HEALTHY, ENVIRONMENTALLY FRIENDLY FOOD OPTION. HERE'S HOW TO GET STARTED PREPARING IT TO EAT.

RIMA KHALEK is an Oakland-based sculptor and chef using unconventional mediums for both arts. She is currently on a road trip across Europe, creating and leaving sculptures along the way.

Many people around the world eat insects **without batting a lash, gagging, or thinking of childhood dares**. They eat insects as snacks or main dishes as an abundant source of protein, healthy fats and minerals. Raising and harvesting insects for human consumption requires considerably fewer resources than livestock does, and releases much less greenhouse gases.

There are more than 1,900 edible insects on Earth. In this article we will focus on three easily accessible ones: crickets, mealworms, and waxworms. We'll use a technique to get them cleaned and cooked that works well with many other insects too. How you prepare them after is completely up to your creative self.

BASIC STARTING POINT

With live insects, it is best to put them in the refrigerator for at least an hour or until you are ready to use them. This will immobilize them by slowing their metabolism, which helps in the collecting and cleaning process. If you prefer, you can place them in the freezer for an hour, which will kill them, and then you can proceed without fear of reanimation.

CRICKETS

Place the crickets (Figure **A**) in the freezer for an hour to eliminate the chance of any escape artists. Then pick them out of their container, being sure to leave any squished ones behind, and clean them by rinsing under cool water.

MEALWORMS

Mealworms (main image on page 36) are the larvae from the mealworm beetle. They have a very earthy flavor. They generally come in a bag with newspaper in it for them to crawl around in.

First, empty the mealworms into a deep bowl, something they can't crawl up and out of, and remove them from the newspaper (Figure **B**).

Next, clean them up. I like to do this while they are still alive and a little slow from the fridge. You will find that there will be plenty of shed skin and a few dead worms, which are easy to find because they are generally darker in color (Figure **C**).

A trick that I found that is a huge help is blowing. Go outside where you can make a bit of a mess, put a handful in a shallow bowl and blow.

TIME REQUIRED:
1–2 Hours

DIFFICULTY:
Easy

COST:
$10–$20

MATERIALS
» **Insects, live.** This article focuses on crickets, mealworms, and waxworms, but many other edible varieties are available.

TOOLS
» **Stove**
» **Cooking pot, large**
» **Refrigerator**
» **Freezer (optional)**
» **Bowl, large**
» **Strainer**
» **Dehydrator**
» **Vacuum sealer**
» **Oven (optional)**
» **Baking sheet (optional)**

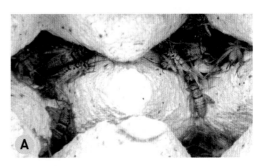
A

B

C

D

E

F

You will blow all their shed skin away. After all the dead and shed skin are separated, give them a nice rinse and stick them in the fridge or freezer for an hour.

WAXWORMS

Waxworms (Figure **D**) are the caterpillar larvae of the wax moth. They have a very neutral flavor and take on what you cook them with.

Waxworks are pretty easy to clean. If they arrive in sawdust, take a large strainer and sift (Figure **E**). Some may squeeze through but it is a pretty painless process of cleaning. Pick out any black dead ones or ones that look like they are on their way out (Figure **F**). Give them a nice rinse and put them back in the fridge for an hour.

COOKING

1. Boil a pot of salted water or prepare stock (savory or sweet). You'll drain the insects quickly, so be ready.

2. Put in your insects. For crickets (Figure **G**), set your timer for 2 minutes. For mealworms (Figure **H**) or waxworms (Figure **I**), 1 minute.

3. When time is up, drain them (Figure **J**), put them in a bowl, and quickly put them in the fridge to cool, about 15 minutes. Don't rinse them.

4. When the insects are cooled, place them on a dehydrator sheet on top of the tray and dehydrate them until they are fully dry but not crunchy (Figures **K** and **L**).

If you are planning to roast them, you can skip the dehydrator and stick them directly onto a baking sheet and into the oven. Set the temperature to 200° F for 2 hours or until desired crispiness.

5. Store your crickets in a mason jar until ready to use; put mealworms and waxworms in a vacuum-sealed bag Figure **M**) or Ziploc bag with air removed. Keep in the refrigerator or freezer until ready to use. Enjoy! ◉

MEAL TIME

Cooking with insects is easier than you might think! It doesn't always have to be whole insects, dried out and crispy. Here are a few recipes you can make right now.

● **MEALWORM FLAPJACKS**
(UK style, not pancakes)
A sweet bar with a savory punch, thanks to mealworms.
eatgrub.co.uk/cook/mealworm-flapjacks

● **MEALWORM MARGARITA**
If you didn't tell your guests, they would probably have no idea their margarita rims were coated in mealworm. You probably should tell them though.
eatgrub.co.uk/cook/mealworm-margarita

● **MEALWORM CHOCOLATE CHIP COOKIES**
A great way to get people to try their first insect-infused meal. Nobody turns down cookies!
allrecipes.com/recipe/260854/mealworm-chocolate-chip-cookies

● **CRICKET FLOUR ALMOND AND GOJI BERRY ENERGY BARS**
Perfect for a snack in your pack. These energy bars use cricket flour for an additional protein punch.
healthyperspective.co/recipes/snacks/cricket-flour-almond-goji-berry-energy-bars-875

● **CRICKET FLOUR BROWNIES**
Just like regular brownies, but made with the protein of the future, crickets.
ediblemichiana.ediblecommunities.com/recipes/cricket-flour-brownies

● **CRICKET FLOUR PIZZA DOUGH**
The world's most popular food can be made with cricket flour as well!
foodrepublic.com/recipes/roasted-tomato-and-pancetta-pizza-with-cricket-flour-dough

—Caleb Kraft

BEST-YET D.I.Y.
COFFEE ROASTER

HACK A FLOUR SIFTER AND A HEAT GUN TO ROAST THE ULTIMATE BEAN

Written and photographed
by Larry Cotton

LARRY COTTON has finally given up on doing anything earthshaking. He loves electronics, music and instruments, computers, birds, his dog, and wife — not necessarily in that order.

TIME REQUIRED:
A Weekend

DIFFICULTY:
Easy

COST:
$50–$90

Over the years, *Make:* has been kind enough to publish three of my DIY coffee roaster builds. (I've promised my editor this will be the last; he's probably skeptical.)

Each roaster has improved on its predecessor in a few ways. This one, based on a simple flour sifter, is my best yet. How is it even better?

- Best heat distribution to every bean
- Beans are totally visible for the entire roast, unlike even some commercial machines
- Quick and easy to dump (and cool) the beans after roasting. Check out sifter roaster videos online for comparison. Hint: They usually skip the critical bean-dumping stage.
- Easily roasts 12oz of green beans — more than some commercial roasters
- Uses cheap parts and materials widely available at Amazon, Harbor Freight, and Lowes.

"But," you say, "I don't roast coffee!" Well, most people don't — yet — but you should seriously consider it. What you may not realize is you're 10–15 minutes away from the freshest coffee obtainable. Green (raw) coffee beans are widely available from many reputable dealers, and of course Amazon. Somebody's gotta roast all those beans and it might as well be you.

Here I'll cover the basic version. You can also make a larger version that quickly cools the beans in a tray with two small fans (both builds are on the project page at makezine.com/go/simple-sifter-coffee-roaster). If you can't decide, you can build the basic roaster on the larger base meant for the fan-cooled version, with the possibility of adding the cooling parts later.

You can make the simple sifter roaster in a weekend, tops, with no special skills, for well under $100. Adding the fans would take a little more time and maybe another 20 bucks.

BUILD YOUR FLOUR SIFTER COFFEE ROASTER

The instructions below require the specific sifter and heat gun (hereafter called HG) in the Materials list. Others will work as well, but will differ in critical details and dimensions.

All unspecified screws are #6×½" Phillips pan-head sheet metal screws.

MATERIALS

- » **Flour rotary sifter, hand cranked, 8 cup** Chef Giant #8541885937, Amazon #B01LYD1TZN, $20
- » **Heat gun, 1500W** Harbor Freight #HF 96289, $15
- » **Gearmotor, 12VDC, 60rpm, 6mm shaft** such as Amazon #B00BX54O8A
- » **Motor shaft coupling, 6mm ID, with flange** Amazon #B07PDYWV3F
- » **"Project" plywood,** ½" thick, about 2'×2'
- » **Aluminum bar or plate,** ¹⁄₁₆" thick × 1½" wide, about 24" total length
- » **Aluminum bar or sheet,** ¹⁄₃₂" or thicker, about 5"×2¼"
- » **Aluminum sheet, .010"–.015" thick, 4"×8"** A scrap of aluminum flashing is ideal; or try an oversize aluminum can
- » **Lumber, 2×4, scrap**
- » **Plywood, ¼", scrap**
- » **Power supply, 12VDC adapter, 1A minimum**
- » **Microswitch, SPST (optional)**
- » **Hookup wire, 18" lengths (2)** 22 gauge preferred
- » **RCA plug and socket (optional)**
- » **Wood dowel, ¼" diameter, 3" long**
- » **Nail, 3", 10D common (~.148" dia.)**
- » **Sheet metal screws, pan head: #6×½" Phillips (1 box)**
- » **Sheet metal screws, #4 or #5 × ¾" (2)** for mounting optional microswitch
- » **Wood screws, flat head, #6×1" Phillips (6)**
- » **Paint or Deft clear wood finish (1 can)** for appearance and weather protection
- » **Machine screws, 6-32:** ¼" (4) and ½" (4) with nuts
- » **Machine screws, 3-0.5mm × 5mm or 6mm (4 or 5)**
- » **Rivets, ⅛"×¼" (optional)** You can use more 6-32×¼" screws and nuts.

TOOLS

Common handheld tools are all that're required to build these roasters. A bandsaw and a drill press would speed things up, of course, and could make the parts more accurate.

- » **Wood saw**
- » **Hacksaw**
- » **Vise**
- » **File**
- » **Drill and bits:** ³⁄₃₂", ⁵⁄₃₂", ⅛", ⁷⁄₆₄", ⁹⁄₆₄" Use a drill press if possible, otherwise a cordless is fine.
- » **Countersink bit**
- » **Thread tap, 6-32**
- » **Measuring tape**
- » **Marking compass**
- » **Pencil**
- » **Screwdrivers**
- » **Soldering iron and solder**
- » **Hot glue gun**
- » **Sander** Use a disk sander if possible, otherwise a hand sander.
- » **Bandsaw or jigsaw (optional)**
- » **Pop rivet tool (optional)**
- » **Stapler (optional)** for fan cooled version

BASES

Top surfaces shown

- Material: ½" "project" plywood

- Countersink all holes from the bottom to accommodate #6 flat-head screws

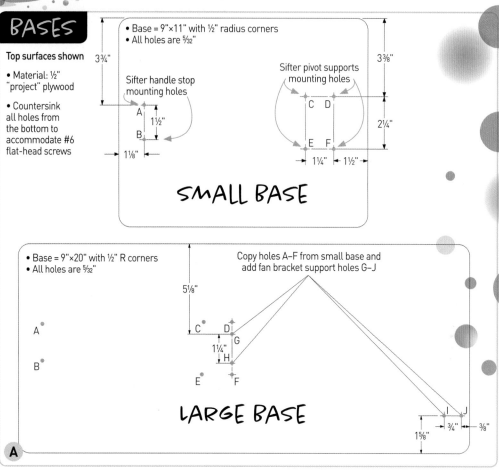

- Base = 9"×11" with ½" radius corners
- All holes are ⁵⁄₃₂"

3¾"

Sifter handle stop mounting holes

A
1½"
B
1⅛"

Sifter pivot supports mounting holes

3⅜"

C   D

E   F

2¼"

1¼"   1½"

SMALL BASE

- Base = 9"×20" with ½" R corners
- All holes are ⁵⁄₃₂"

Copy holes A–F from small base and add fan bracket support holes G–J

5⅛"

A

B

C   D
G
1¼"
H
E   F

I   J
¾"   ⅜"
1⅝"

LARGE BASE

A

SUPPORTS AND STOPS

Except as noted, make one each from ½" ("project") plywood

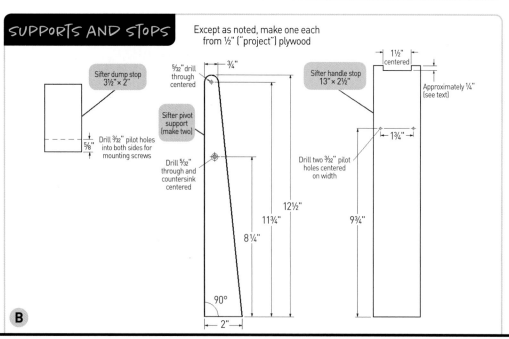

Sifter dump stop
3½" × 2"

Drill ³⁄₃₂" pilot holes into both sides for mounting screws

⅝"

⁵⁄₃₂" drill through centered

¾"

Sifter pivot support
(make two)

Drill ⁵⁄₃₂" through and countersink centered

12½"
11¾"
8¼"

90°

2"

1½" centered

Sifter handle stop
13" × 2½"

Approximately ¼" (see text)

1¾"

Drill two ³⁄₃₂" pilot holes centered on width

9¾"

B

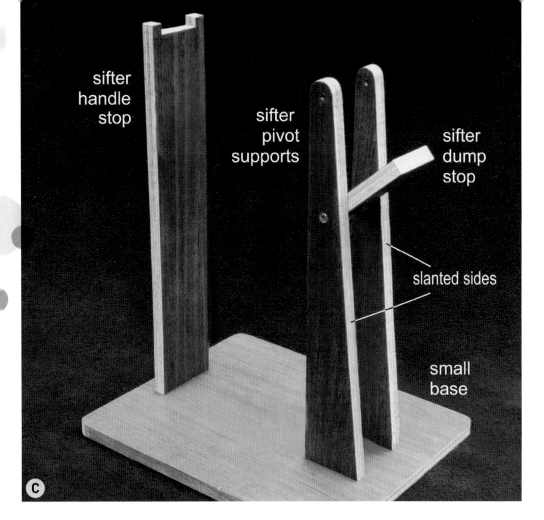

sifter handle stop

sifter pivot supports

sifter dump stop

slanted sides

small base

C

1. MAKE THE BASE

1a. Cut a base from ½" plywood. Two sizes (for both versions) are shown in Figure **A**. Drill and countersink the 6 holes labeled A–F for various supports and stops (or the 10 holes A–J if you're making the large base).

1b. Make the plywood supports and stops as shown in Figure **B**. Later, you'll drill all these parts from one end to match their respective base mounting holes.

1c. Paint the base, supports, and stops to suit, or leave the wood natural and spray with clear Deft.

1d. Center the sifter pivot supports and handle stop above their corresponding screw holes in the base and drill ³⁄₃₂" pilot holes into them from the bottom. Mount them with #6 ×1" Phillips flat-head wood screws. Note the positions of the slanted

sides of the pivot supports (Figure **C**).

Mount the sifter dump stop between the two pivot supports at about a 45° angle with a #6 ×1" pan-head sheet metal screw from each side.

2. MODIFY THE FLOUR SIFTER

2a. Remove the sifter's agitator assembly as follows:

1. Remove the brass nut on the end of the hand-crank axle. Don't lose it; you'll use it again.
2. Discard* the nylon washer.
3. Hold the agitator while turning the axle crank counterclockwise to unscrew the axle.
4. Pull the axle out and discard* the agitator. I've seen both threaded and unthreaded axles; either works fine.

*Actually, never discard anything. You may need it for your next project!

MISCELLANEOUS WOOD AND ALUMINUM PARTS

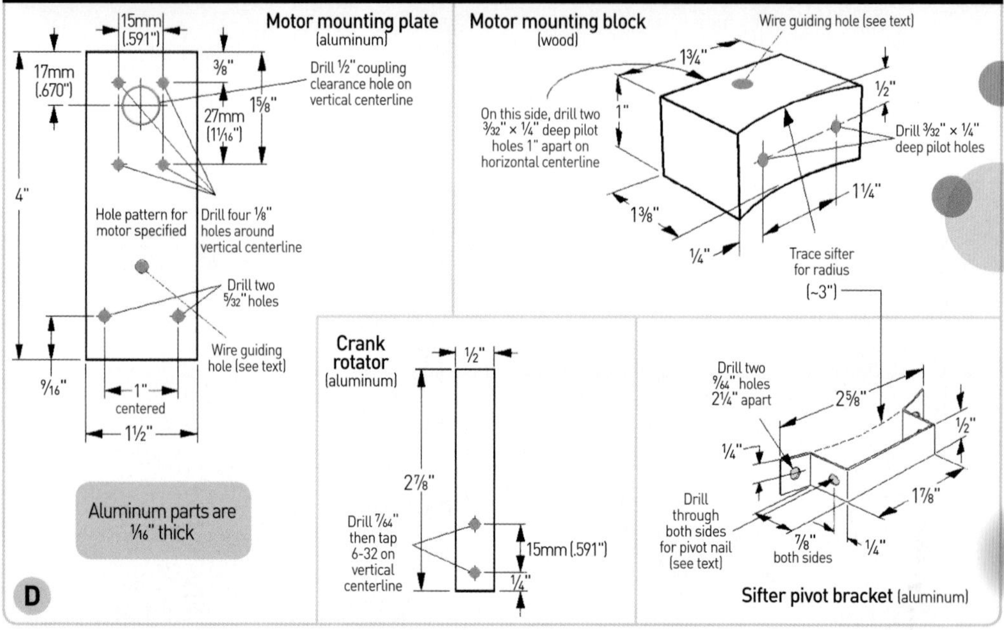
Motor mounting plate (aluminum)
15mm (.591")
17mm (.670")
3/8"
1 5/8"
27mm (1 1/16")
Drill 1/2" coupling clearance hole on vertical centerline
4"
Hole pattern for motor specified
Drill four 1/8" holes around vertical centerline
Drill two 5/32" holes
Wire guiding hole (see text)
9/16"
1" centered
1 1/2"
Aluminum parts are 1/16" thick
D
Motor mounting block (wood)
Wire guiding hole (see text)
1 3/4"
1/2"
On this side, drill two 3/32" × 1/4" deep pilot holes 1" apart on horizontal centerline
1"
Drill 3/32" × 1/4" deep pilot holes
1 1/4"
1 3/8"
1/4"
Trace sifter for radius (~3")
Crank rotator (aluminum)
1/2"
2 7/8"
Drill 7/64" then tap 6-32 on vertical centerline
15mm (.591")
1/4"
Drill two 9/64" holes 2 1/4" apart
2 5/8"
1/2"
1/4"
Drill through both sides for pivot nail (see text)
1 7/8"
7/8" both sides
1/4"
Sifter pivot bracket (aluminum)

2b. Remove the crank's black plastic knob. You can try breaking it with a hammer, but I resorted to sawing it off with a hacksaw.

2c. Holding the axle crank in a vise, hacksaw the end to leave about ½". File the end of the crank. (Figure J on the opposite page shows this as it will be assembled later.)

2d. Re-insert the axle into the sifter as it was (crank on the right with sifter handle toward you) and securely replace the nut. Ensure the axle still rotates freely in the sifter. You'll make and attach two bean-stirring paddles to it later.

3. MAKE WOOD AND ALUMINUM PARTS

3a. Follow Figure D to make all parts in Step 3. Make the motor mount from a scrap of 2×4 wood. Paint it for appearance only; I like red!

3b. Make the motor mounting plate. The four-hole pattern (15mm × 27mm) for the 3mm motor-mounting machine screws must match the screw holes in the specified motor. The plate holes can be drilled slightly oversize if necessary for alignment.

3c. Drill wire guiding holes in the motor mount and the motor mounting plate. Start with the ⅛" holes shown in Figures D and G; details later.

3d. Make the crank rotator as shown.

3e. Cut another ½" aluminum strip about 6" long and make the sifter pivot bracket. I used a 3" 10D common nail (.148" dia.) for the pivot itself, and drilled 5/32" holes to just clear the nail.

I'm not much of a sheet-metal fabricator, so I drilled the bracket using the technique shown in Figure E, which ensures both pivot holes are in line with each other. Then I finished drilling and bending before cutting the bracket to length. You may know a better way to do this!

4. MOTORIZE THE FLOUR SIFTER

4a. Attach the 12VDC motor to its mounting plate with four 3-.50 (metric thread) × 5mm or 6mm long machine screws.

4b. Attach that assembly to the wood motor mount with two screws.

4c. Attach the motor coupling to the crank rotator

E

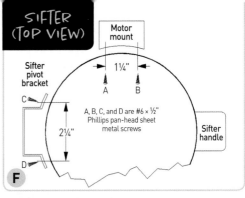

SIFTER
(TOP VIEW)

Motor mount

Sifter pivot bracket

1¼"

A   B

A, B, C, and D are #6 × ½" Phillips pan-head sheet metal screws

C

2¼"

D

Sifter handle

F

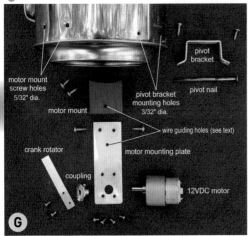

motor mount screw holes 5/32" dia.

pivot bracket

pivot nail

pivot bracket mounting holes 3/32" dia.

motor mount

wire guiding holes (see text)

crank rotator

motor mounting plate

coupling

12VDC motor

G

sifter handle

motor assembly

pivot bracket

1/2"   1/4"
to screw holes

H

using two of the 6-32 × ¼" Phillips pan-head machine screws.

4d. Slip that assembly onto the motor shaft and tighten the coupling setscrew on the flat surface of the shaft. If a screw didn't come with the coupling, use a 3-.50 × 5mm or 6mm machine screw.

4e. Figure **F** illustrates the top of the sifter. Note the positions of the motor mount and pivot bracket relative to the sifter handle.

Drill two ³⁄₃₂" holes for the pivot bracket screws and two ⁵⁄₃₂" holes for the motor mount screws as shown in Figures **G** and **H**.

A drill press makes this easier (Figure **I**). Otherwise use an electric drill or screwdriver with new drill bits and a few drops of oil; that steel is pretty tough.

4f. Mount the five-piece assembly to the sifter using two screws from the inside. The sifter crank and its rotator should now intersect (Figure **J**) and, when rotated, clear the wood motor mount so that the sifter axle rotates freely. (The motor will

I

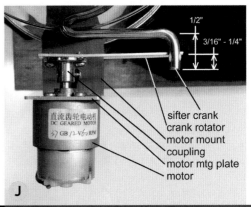

1/2"

3/16" - 1/4"

直流齿轮电动机
DC GEARED MOTOR
GB /2-V6uRPM

sifter crank
crank rotator
motor mount
coupling
motor mtg plate
motor

J

pivot bracket in
pivot supports

10D x 3"
common
nail

K

**WIND-BREAK FUNNEL
BEFORE FORMING**

5°
(total angle 185°)

$3\frac{7}{8}$" R*
(4" max.)

Make from aluminum
flashing or other thin
sheet metal

$1\frac{5}{8}$" R

Drill four holes
to suit (see text)

*For any sifter other than the one specified,
use $4\frac{1}{4}$" outside radius. Trim as needed.

L

wind-break funnel

two
rivets or
6-32 machine
screws and nuts

M

offer resistance, but it's OK to rotate its shaft by hand to check the clearance.)

4g. Attach the pivot bracket to the sifter using two screws from the outside. Alternately, enlarge those mounting holes to $\frac{5}{32}$" and use two more $\frac{1}{4}$"-long 6-32 machine screws and nuts.

4h. Mount the entire sifter assembly atop the two sifter pivot supports using the 10D × 3" nail (Figure **K**), hacksawed and filed to length if desired (although the point could help in assembly).

Ensure the sifter handle fits into its recess in the top of the sifter handle stop, and that the sifter pivots smoothly from bean-load (sifter vertical) to bean-dump positions. Depending on your machining and assembly accuracy, you may have to adjust the location of the stop or one of the pivot supports.

5. MAKE HEAT GUN MOUNTING PARTS

5a. Make the wind-break funnel to Figures **L** and **M** . Wrap the C-shaped aluminum onto itself, with about a $\frac{1}{4}$" overlap. Test to see that its smaller end just clears the nozzle of the specified HG. Drill two small aligned holes in each end and join them with either two $\frac{1}{8}$" dia. × $\frac{1}{4}$" long rivets or 6-32 machine screws and nuts. Slip the smaller end over the HG nozzle. If it's too small, trim a bit off the funnel; if it's too big, it'll still work, but you could "ellipsize" the hole somewhat for a better fit.

5b. Make the HG nozzle bracket from sheet aluminum ($\frac{1}{32}$" minimum) to Figure **N** . I cut the $1\frac{1}{2}$" hole with an adjustable bit in a drill press, but you could also jigsaw it —carefully! — after drilling a $\frac{3}{8}$" hole to clear the jigsaw blade. Fasten it to the sifter handle stop with two screws, protruding straight out at 90° (Figure **O**).

5c. Make the HG locator from $\frac{1}{4}$" plywood to Figure **P** . Hold it centered on the air intake cap of the HG, while gluing 3 wood pieces ($\frac{1}{2}$"×$\frac{1}{2}$"×$\frac{1}{4}$" thick) to the ends of the locator arms, adjacent to the cap. You can see these in Figures **R** and **S** .

5d. Put the HG, its locator, the nozzle bracket, funnel, and sifter in their roasting positions, then

BEST DEAL

Subscribe today and get Make: delivered right to your door for just $39.99! You **SAVE 33% OFF** the cover price!

Name _____ (please print)

Address/Apt. _____

City/State/Zip _____

Country _____

Email (required for order confirmation) _____

☑ **One Year**
$39.99   ☐ **Payment Enclosed**   ☐ **Bill Me Later**

B9MN53

Make:

Make: currently publishes 4 issues annually. Allow 4-6 weeks for delivery of your first issue. For Canada, add $9 US funds only. For orders outside the US and Canada, add $15 US funds only.

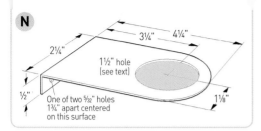

N

2¼" 3¼" 4¼"

1½" hole
(see text)

½"   One of two ⁵⁄₃₂" holes
1¾" apart centered
on this surface

1⅛"

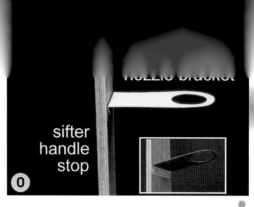

sifter
handle
stop

O

HEAT GUN LOCATOR

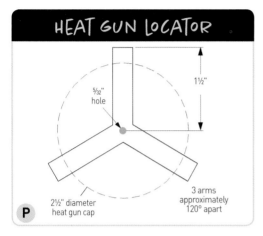

⁵⁄₃₂"
hole

1½"

2½" diameter
heat gun cap

3 arms
approximately
120° apart

P

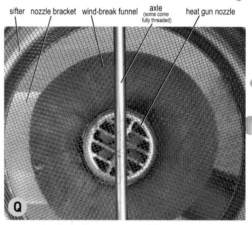

sifter nozzle bracket wind-break funnel   axle (some come fully threaded)   heat gun nozzle

Q

look straight down through the sifter and funnel into the HG's nozzle (Figure Q).

5e. Position the HG locator to hold the HG centered under the sifter. Then, pencil-mark the locator's position on the base. Ensure the cap stays off the base for good airflow (Figure R).

5f. Remove the HG and mount the locator to the base with one screw (Figure S).

5g. Cut a 3" piece of ¼" wood dowel to trap the HG handle against a pivot support. Trace around where the dowel touches the base. Drill a ¼" hole and insert the dowel (Figure T). To allow for possible servicing of the HG, don't glue it in.

6. MAKE THE PADDLES

The main goal of this — and every — coffee roaster is to expose all the beans to the same amount of heat for the same amount of time. Although the stock flour sifter agitators work OK, twisted paddles are better: the beans are robustly stirred in alternate directions while being evenly heated.

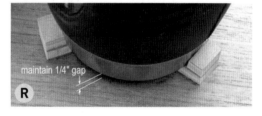

maintain 1/4" gap

R

S

T

6a. Make two paddles to Figure 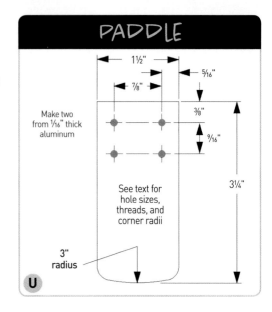 **U**. Cut one end of both to about a 3" radius, approximately matching the sifter screen. Round the corners somewhat so the paddles don't drag the screen. Mark where the holes will be but don't drill the holes yet.

6b. Overlap the square ends of the paddles, and clamp them together for a total length of 5³⁄₁₆". Drill four ⁷⁄₆₄" holes through both paddles at the same time using the dimensions in Figure **U**.

6c. Unclamp the paddles and tap only one (either) paddle's holes to 6-32 machine-screw threads. We won't be using nuts here.

6d. Enlarge the holes in the other paddle to ⁵⁄₃₂" to clear 6-32 machine screws.

6e. For optimal bean circulation, twist the paddles, using a ¾"×2"×8" scrap of wood. Cut a slot 1½" into the center of one end, to just clear the paddle's thickness.

6f. Clamp the drilled end of one paddle in a vise until about 2" sticks out. Slide your paddle twister over the protruding end and give it a slow but strong twist in either direction (Figure **V**). The angle's not particularly critical, but shoot for 25°–30°.

6g. Do the same for the other paddle except twist it in the opposite direction!

6h. Roughly position the paddles on both sides of the sifter axle with their screw holes lined up. Lightly clamp them together on the axle with four 6-32×½" Phillips-head machine screws, two on

PADDLE

Make two from ¹⁄₁₆" thick aluminum

1½"
⁷⁄₈"
⁵⁄₁₆"
³⁄₈"
⁹⁄₁₆"
3¼"

See text for hole sizes, threads, and corner radii

3" radius

U

each side of the axle (Figure **W**).

6i. Hand-spin the crank to check for gaps between the ends of the paddles and the sifter screen. Your goal is to have a small gap, ¹⁄₁₆"–⅛", for each paddle. It's OK for one paddle, but not both, to lightly drag the sifter screen. If necessary, loosen the screws, shift the paddles a bit, retighten the screws, and retest. If both paddles insist on dragging the screen, take the assembly apart, trim a bit from one or both paddles' curved ends, remount them to the axle and test.

WARNING: Paddle-to-screen drag reduces the paddles' bean-stirring strength and may wear a hole in the screen. Also, beans could become trapped between the ends of the paddles and the joint between screen and sifter, stalling the motor and if left that way, ruining the roast — yikes!

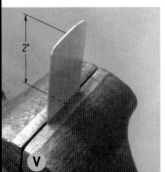

2"

V

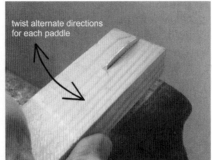

twist alternate directions for each paddle

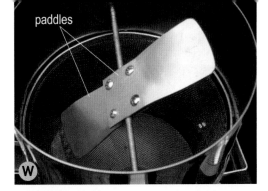

paddles

W

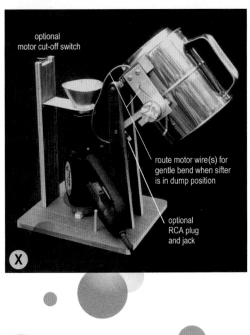

optional
motor cut-off switch

route motor wire(s) for
gentle bend when sifter
is in dump position

optional
RCA plug
and jack

X

7. WIRE THE PADDLE MOTOR

The specified motor runs on 12VDC. Depending on the length and diameter of the power supply (adapter) cord, you may be able to just cut and route it to the paddle motor through the ⅛" holes drilled in the motor mount and mounting plate. Enlarge the holes if necessary to clear the cable.

I added an RCA plug and jack, mostly for troubleshooting, shown under the sifter dump stop in Figure **X**. The plug-to-motor distance should be ~18". In any case, the wires or cable to the motor must be flexible and long enough to be subjected to repeated (but gentle) bending when dumping beans.

I also added a microswitch (SPST, normally-off) to the top of the sifter handle stop (Figures O and X), so when the sifter handle is lifted off its stop, the paddles will stop rotating. This isn't critical, but may prevent kicking a few beans to the ground/floor/patio while loading them. Use two #4 or #5 × ¾" pan-head sheet metal screws to mount the switch. You may have to enlarge the switch's mounting holes with a ⅛" drill bit. Wire it in series with either of the wires to the sifter motor (Figure **Y**).

However you wire, try to hide the wires as much as possible. Strip and solder both wires to the motor. Mine is wired so that the paddles turn counterclockwise when viewing them from the motor side. The roaster should work either way; just reverse the motor wires. If you notice a speed difference in clockwise vs. counterclockwise revolution, wire for higher RPMs.

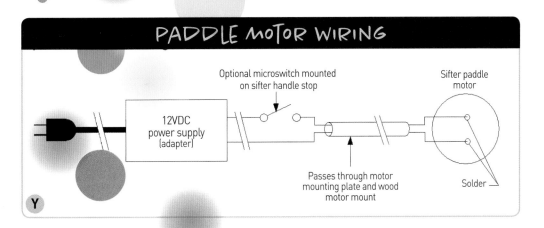

PADDLE MOTOR WIRING

Optional microswitch mounted
on sifter handle stop

Sifter paddle
motor

12VDC
power supply
(adapter)

Passes through motor
mounting plate and wood
motor mount

Solder

Y

USING YOUR ROASTER

Coffee roasters without smoke and chaff removal systems — like this one — should be used in a dry, well-ventilated place, preferably outdoors. Always use the wind-break funnel. I've roasted coffee in 40°F temps with no problems. Stay with the roaster from start to finish. Beans darken quickly at the end of the roast. Keep kids at a distance and use plain old common sense. To get started:

1. Before adding any beans to the sifter, plug the paddle motor adapter, heat gun, and fans (if you built the cooling-tray version) into a dedicated 15A circuit. You can add up to 25 feet of 14-gauge extension cord, if necessary. Read the Harbor Freight heat gun manual. Note that its switch has high and low heat settings but doesn't have a "cool" setting. Leave the heat gun off for now.

2. If you built the larger version, place the tray over the cooling fans; they must turn on when the tray is present.

3. If you added a microswitch to the top of the sifter handle stop, it must turn the paddle motor on when the handle is in place.

4. For your first batch, with the heat gun off, slowly pour 1 cup of green coffee beans into the sifter. If you installed the switch on the sifter handle support, just lift the sifter a bit to stop the paddles.

The beans should move smoothly and vigorously. If not, you may have too large a gap between the end of one or both paddles and the sifter screen. If the paddles are too long, an occasional bean could get caught between a paddle and the sifter screen anchoring rim, stalling the motor. If this happens, try running the motor the other way. Otherwise, unplug everything and revisit Step 6i.

5. Switch the heat gun on. Use the low setting if the ambient temperature is above 60°F, and the high setting for cooler temps. Beware the hot nozzle!

6. After a few minutes, the beans should begin to turn yellow and shed some chaff. They'll gradually darken from tan to brown, and you'll hear them crackle (aficionados call that the "first crack") as they expand and release moisture. You might want to blow chaff from the sifter occasionally.

7. If after 5 minutes the beans have not changed color, switch your heat gun to the high setting.

8. Beans should fully roast in 12–15 minutes. When they're roasted to your liking and/or just starting to crack again ("second crack"), turn the heat gun off. Some bluish smoke can be expected. Some people like to roast well into the smoke stage; I do not.

9. If you built the basic roaster, quickly dump the roasted beans into a large baking pan. It acts as a heat sink, minimizing the beans roasting further.

> **CAUTION:** Be very careful dumping the beans; they are extremely hot!

10. I've roasted up to 1½ cups of green beans per batch with no problems. Experiment!

11. Purists say to wait a day before grinding and brewing to allow the beans to out-gas. Good luck with that!

GOING FURTHER

Feeling ambitious? Check the project page for a video of my unnecessarily complicated, semi-automatic version (below). The beans are auger-fed into the sifter at the push of a button; after roasting, the sifter dumps them automatically and returns for more. This one's obviously way too silly and complex for its own article — or *is* it? — but it's a hoot to watch. ◔

 For more roaster photos and videos, and the extended build with cooling fans, visit makezine.com/projects/ simple-sifter-coffee-roaster.

LARRY'S DIY ROASTER SAGA
is now complete! ...we hope.

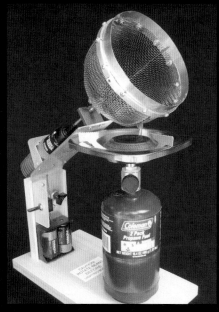

CAMP STOVE COFFEE ROASTER
Make: Vol. 08
makezine.com/projects/coffee-roaster-the-nirvana-machine

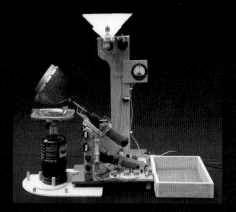

SEMI-AUTOMATIC COFFEE ROASTER
Make: Vol. 46
makezine.com/projects/semi-automatic-coffee-roaster

DOG BOWL COFFEE ROASTER
Make: Vol. 65
makezine.com/projects/dog-bowl-coffee-roaster

HELLO DRINKBOT!

USE A RASPBERRY PI TO BUILD THE "HELLO WORLD" OF COCKTAIL ROBOTICS

Written and photographed by Rich Gibson

Cocktail robotics is an engaging cross between engineering, art, and gastronomy that provides many entry points for making, craft, performance, and exploring hedonistic interfaces between technology and humanity.

Cocktail robots don't need to serve alcohol, and they don't need to be electronic. There are a lot of fun "cocktail robots" that dispense other drinks, and robots that dispense cocktails without a computer (see "The Corpse Reviver," page 59).

But the goal of the Hello Drinkbot project is to get people making more cocktail robots by providing a simple and affordable solution to the "hello world" problem of "How do I dispense cocktails under computer control?"

Hello Drinkbot aims to make it easier to get started, to lower the bar of entry so that more people can enjoy the rewards of this particular obsession. To paraphrase artificial intelligence pioneer Danny Hillis, we would like to provide a platform which tolerates the beginner's skill and limited time, while rewarding more attention and creativity. This is "hello world," so imperfections should perhaps be embraced!

This open source project includes plans and software to make a four-ingredient cocktail-dispensing (or Italian soda-dispensing) drinkbot, expandable to eight ingredients, as well as other resources and code you can use to customize, explore, and expand on the basic model.

I presented Hello Drinkbot as a hands-on activity at Maker Faire Bay Area 2019. The workshop was a lot of fun, and many lessons were learned. On the plus side, I think that all the robots went home able to dispense liquids. Yay!

Since then I've worked on tweaks, improvements, patches, and bug fixes. Hello Drinkbot is ready to start pouring for you.

RICH GIBSON works at OpenFiber on mapping and databases. He worked on the Gigapan and Explorable Microscopy projects for Carnegie Mellon University and NASA's Intelligent Robotics group, and coauthored *Mapping Hacks* and *Google Maps Hacks* for O'Reilly.

TIME REQUIRED:
A Weekend

DIFFICULTY:
Intermediate

COST:
$100–$150

MATERIALS
» **Raspberry Pi 3 B+ single-board computer** Older Pi's will work, but you need to get the right headers to fit on the Motor Hat (see below). A Pi Zero W also appears to work, but the mounting holes are slightly different.
» **MicroSD card, 32GB, with adapter** 8GB is probably more than enough, but maybe you want to do more cool things with your bot.
» **Adafruit DC & Stepper Motor Hat for Raspberry Pi** Adafruit #2348. You'll solder the headers and screw terminal blocks.
» **Peristaltic liquid pumps, 12VDC, 3mm ID × 5mm OD (4–8)** I used Intllab #606015745006, Amazon #B0791YL351. Four are enough for a basic bot, but this project can drive up to 8 pumps.
» **Silicone tubing, food grade, 3mm ID × 5mm OD, 2–3 meters total length** such as Amazon #B0799FKWQ9
» **Plywood or acrylic, ¼" (6mm) thick, about 2'×4' total area (optional)** for the laser-cut case. Alternatively, you can just use a hole saw to mount the pumps in the box of your choice.
» **Touchscreen display for Raspberry Pi, 7" (optional)** Raspberry Pi #7TOUCH, Amazon #B0153R2A9I. So far, Hello Drinkbot doesn't have a case that lets you cleanly mount the touchscreen. If you use a touchscreen or external monitor, you may also want a keyboard. I use a Bluetooth keyboard, Amazon #B014EUQOGK; it includes a trackpad, and has worked well so far.
» **Hex standoffs, brass, M2.5, 11mm body, 6mm screw (6)** for Raspberry Pi, Amazon #B07KM27KC6
» **Machine screws, stainless steel: M2.5×12mm (4) and M3×12mm (8–16)**
» **Hex nuts, stainless steel, M3 (8–16)**
» **Power supply, 5V 2.5A Micro USB** for the Raspberry Pi such as Amazon #B075VP7KKM
» **Power supply, 12V 3A, with 5.5mm/2.1mm barrel plug** for the pumps, such as Amazon #B07CWVFGNN
» **Power pigtail, 12V, with 5.5mm/2.1mm barrel jack** such as Amazon #B072BXB2Y8
» **Crimp terminals, female, 18–22 gauge (8–16)** Amazon #B01962MW2G
» **Hookup wire, 18 gauge**

TOOLS
» **Laser cutter (optional)** Cut the files yourself or send them out to a cutting service. Or just mount this project in a different box of your choice.
» **Wood glue**
» **Soldering iron and solder** to solder the headers on the Adafruit Motor Hat
» **Screwdrivers**
» **Pliers** for crimp connectors

Adobe Stock - evgeniya_m

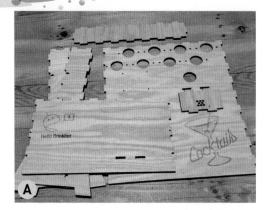

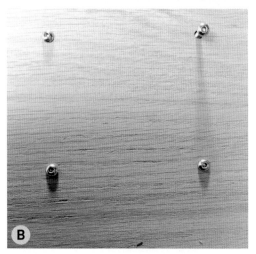

BUILD YOUR HELLO DRINKBOT

If you have a better way to do any of this, please let me know. I made it up as I went along.

1. ASSEMBLE THE CASE

If you're building the laser-cut case, download the *.dxf* or *.svg* file at github.com/RichGibson/hellodrinkbot/tree/master/hardware/stand_alone and cut the case from 6mm (¼") material. This is a stand-alone case with holes cut for 8 peristaltic pumps, Pi mounting screws, and tubing management. It has six parts: 2 sides, 1 back (with the cutouts for pumps and Pi screws), 1 front, 1 top, and 1 dispenser/tubing management shelf (Figure A).

You can glue the case together or use four M3×12mm screws and nuts to fasten it if you want to break it down later. For now you can leave the front off.

2. MOUNT THE PI

Insert four M2.5×12mm screws through the back for the Pi mounting (Figure B). Then screw four standoffs onto those screws (Figure C). Hand tight is fine.

Put the Raspberry Pi on those screws, and secure it with two M2.5 nuts on the side of the Pi with the GPIO header. Screw two more standoffs onto the standoffs on the other side of the board (Figure D).

Solder the headers onto the Adafruit motor hat, and then attach it onto the Pi's GPIO headers. Use two M2.5 nuts to secure it (Figure E).

3. MOUNT THE PUMPS

Put the pumps through the holes, and attach each one using two M3×12mm screws with nuts on the inside of the case (Figure F). Figure G shows the rear side of the bot.

4. WIRE THE PUMPS

A bit of trivial trickery allows us to double the number of pumps that our motor hat can control. The motor board has 4 full H-bridges which allow it to control four DC motors in forward or reverse. But we don't normally need reverse, so we're splitting each H-bridge into two half H-bridges,

D

E

F

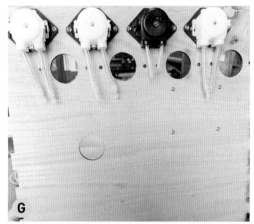

G

allowing us to control 8 motors. In the software we'll move **FORWARD** to control motor 1, and **BACKWARD** to control motor 2.

Connect the black ground wire harness to each motor, either the top or bottom lug, just be consistent (Figure **H**). You can solder to the motors or use crimp connectors. Route the wires in a way that is pleasing to you, cable tie them if you would like. Clip them shorter to fit your desire for order.

On the Motor Hat, connect the black ground wire to either of the screw terminals labeled GND in the left group of screw terminals. Connect the first motor to the first pin of M1, and the second motor to the second pin of M1. The third motor to pin 1 of M2, and the fourth motor to pin 2 of M2. If you have 8 motors, continue that pattern on M3 and M4.

Figure **I** on the following page shows the wiring completed. Tidy enough for me.

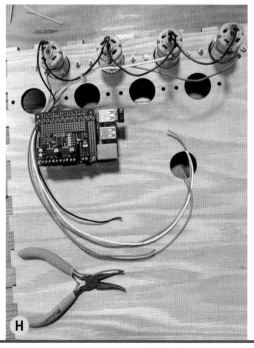

H

5. CONNECT POWER

Connect the red wire of the 12-volt pigtail to + on the motor hat, and the black wire to – (Figure **J**). Route the power pigtail out the hose hole in the case back. For strain relief, cable-tie the pigtail to one of the standoffs on the Pi.

The insides are complete. Make sure your configured SD card is inserted (see "Set Up the Software" below to configure). Connect the 5V 2.5A Micro USB power supply to the Pi, and the 12V power supply to the 12V pigtail (Figure **K**).

6. CONNECT THE DISPENSER

The dispenser board is mounted on front face of the case back (Figure **L**).

Connect lengths of silicone hoses to your pumps, and run them into the dispenser holes (Figure **M**). Once you've got everything working, you can zip-tie them together to keep them tidy.

SET UP THE SOFTWARE

Hello Drinkbot can be operated by my forked version of the open source Bartendro software that's been kindly shared by Party Robotics. I've also written some utilities that let you control Hello Drinkbot directly via Python commands.

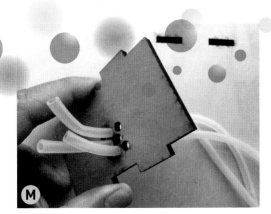

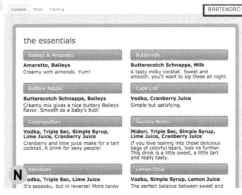

Cocktails Shots Trending

BARTENDRO

the essentials

Baileys & Amaretto

Amaretto, Baileys
Creamy with almonds. Yum!

Buttermilk

Butterscotch Schnapps, Milk
A tasty milky cocktail. Sweet and smooth, you'll want to sip these all night.

Buttery Nipple

Butterscotch Schnapps, Baileys
Creamy mix gives a nice buttery Baileys flavor. Smooth as a baby's butt!

Cape Cod

Vodka, Cranberry Juice
Simple but satisfying.

Cosmopolitan

Vodka, Triple Sec, Simple Syrup, Lime Juice, Cranberry Juice
Cranberry and lime juice make for a tart cocktail. A drink for sexy people!

Gummy Bears

Midori, Triple Sec, Simple Syrup, Lime Juice, Cranberry Juice
If you love tearing into those delicious bags of colorful bears, look no further. This drink is a little sweet, a little tart and really tasty.

Kamikaze

odka, Triple Sec, Lime Juice
It's seppoku, but in reverse! More tangy

Lemon Drop

Vodka, Simple Syrup, Lemon Juice
The perfect balance between sweet and

This project assumes you've got a working Raspberry Pi — to configure a brand-new Pi, follow the guide at raspberrypi.org/help.

1. DOWNLOAD THE CODE

Check the project page at hellodrinkbot.com to download a new disk image (or to get instructions for installing the software on an existing Pi). This includes the Hello Drinkbot code from github.com/RichGibson/hellodrinkbot, the Python library for the Adafruit Motor Hat, my fork of the Bartendro code from github.com/RichGibson/bartendro, and an installation of Bruce Wilcox's ChatScript.

To burn the disk image to your SD card, follow the instructions at raspberrypi.org/documentation/installation/installing-images. Then insert the card into the Pi's microSD slot.

2. ACCESS THE RASPBERRY PI

You can either connect a monitor, keyboard, and mouse to your Pi, or you can access it from another computer via SSH. Putty is an SSH client for Windows; on a Mac you can use Terminal.

If you connect to the Pi via wireless, you can SSH to the **wlan0** interface. The username is **pi**, the password is **raspberry**. (Yes, leaving the default username and password is a security risk. You can change it if you want.) To connect, type the command:

```
ssh pi@10.0.0.1
```

You can also connect the Pi to your computer or hub via Ethernet cable. If you have a Mac, enable Internet Sharing (System Preferences→Internet Sharing). Your Pi will probably get the first address, 192.168.2.2. To connect, type:

```
ssh pi@192.168.2.2
```

The Pi is set up with Multicast Domain Name Service (mDNS), so you should also be able to connect to it as *<hostname>.local*. The hostname is set in */etc/hostname* to **hellodrinkbot**. Type:

```
ssh pi@hellodrinkbot.local
```

The SSID, the access point name, is set in */etc/hostapd/hostapd.conf*.

3. TEST FROM THE COMMAND LINE

Hello Drinkbot code includes utility programs demonstrating how to talk to the motor control hat from Python. Connect to your Pi, then type:

```
cd hellodrinkbot/software/utility
python ./pumptest.py 2
```

This will test all the pumps, turning each on for 2 seconds, and then moving to the next. Your drinkbot is alive! Now you can start commanding it from your own Python programs, or you can get the party started with Bartendro.

4. USE BARTENDRO

Bartendro will offer you a menu of drinks that can be made with your chosen ingredients (Figure 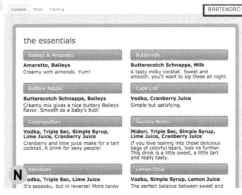). It also lets you dispense individual shots, customize ingredient amounts, add your favorite boozes, and create your own recipes.

Web interface: Associate with the "HelloDrinkbot" access point and then go to hellodrinkbot.local or 10.0.0.1 in your browser. You'll see the Bartendro graphical interface in your browser, and you're good to go.

RESTful interface: I tweaked Bartendro so that you can also make drinks by calling the RESTful interface from your own code. For example: http://hellodrinkbot.local/ws/drink/12?booze14=75&booze11=75

This way you can automate cocktails from any web page, award them as prizes in games, voice-command them from a smart speaker ... use your imagination!

5. CUSTOMIZE BARTENDRO

To access the Bartendro admin pages, click on the Bartendro logo in the upper right corner of the screen. Default username is **bartendro**, password **boozemeup**. Here's where you specify your Dispensers and customize your Booze list, Drinks recipes, and more. Exit admin by clicking on Menu on the upper left.

My fork of Bartendro has a new feature: the graphical Shots menu now shows pictures of the different boozes. You can update these images if you like. SSH to your robot, then type:

```
cd bartendro ui
sqlite3 bartendro.db
```

Identify your booze:

```
select id, name, image from booze where
name like 'Vod%';
1|Vodka|
55|Vodka, Blavad Black |
58|Vodka, Bacon Infused|
```

Then update the picture:

```
sqlite> update booze set image='potato.
jpg' where id in (1,55,58);
```

Put the pictures in *bartendro/ui/content/static/images/*. (Really there should be a way to add pictures at the Booze admin screen. Patches welcome!)

6. LOGS AND DEBUGGING

You can look at the logs by SSH'ing into the Pi over any interface (10.0.0.1, 192.168.2.2, or hellodrinkbot.local). System logs are in */var/log* and web and Bartendro logs are in */var/log/nginx*.

The following commands may be useful:

```
tail -f /var/log/nginx/bartendro-error.log
```

```
tail -f /var/log/nginx/bartendro-combined.
log
tail -f /var/log/nginx/access.log
tail -f /var/log/nginx/error.log
tail -f /var/log/syslog
```

GET YOUR DRINKBOT ON

I brought a Hello Drinkbot to an open house at Chimera Arts & Maker Space in Sebastopol, California (Figure **O**), and was able to serve a bunch of happy people Mudslides, White Russians, and Irish Russians (and variants on those). I used the modified Bartendro software, and aside from a little off-by one error (now fixed), everything worked great.

SOFT DRINKS TOO

At another party recently, I loaded a Hello Drinkbot with various flavors of Torani syrup and set the shot size to 10ml. My grandson Finn took over bartending duties and enjoyed dispensing Italian sodas for people (Figure **P**).

The peristaltic pumps can't dispense carbonated liquid, so people would put their own ice and carbonated water into their cups and hand them to Finn. He used the Shots menu in Bartendro. It shows a list of the loaded ingredients and lets you dispense shots individually from those ingredients.

CLEANING YOUR BOT

After I'm finished dispensing, I run water through all the pumps, and then I run some vodka through.

GOING FURTHER

Your bot has a version of ChatScript installed — so you can talk to it like a real bartender. ChatScript is an open source chat engine written by Bruce Wilcox (brilligunderstanding.com). The Wikipedia page has a good description, and ChatScript's official pages and Github repository (github.com/ChatScript/ChatScript) are filled with fascinating documentation and explorations into the issues involved with using natural language.

The article "How to build your first chatbot using ChatScript" has a nice introduction (freecodecamp.org/news/chatscript-for-

O

P

beginners-chatbots-developers-c58bb591da8).
But here's a shorter version. Connect to your Pi,
then type:

```
cd Chatscript
BINARIES/LinuxChatScript64 local
```

Enter your name when prompted, or any
identifier. ChatScript maintains persistence with
a user across sessions.

Interact with the bot. After a couple of
questions, it will ask if you want a cocktail. If you
say yes, it will pour a tiny cocktail!

There are multiple ways you can interface
ChatScript with your bot. I scripted mine to chat
in character as The Dude from *The Big Lebowski*,
then pour me a White Russian. ⊘

HELLO DRINKBOT RESOURCES:

Project page: hellodrinkbot.com

Facebook group: facebook.com/
groups/602734573508700

Github repos: github.com/RichGibson/
hellodrinkbot and github.com/
RichGibson/bartendro.

MORE COCKTAIL BOTS

The Corpse Reviver is a beautiful, one-
of-a-kind, hand-operated, all-mechanical
cocktail mixer built by Benjamin Cowden
for the 2009 Roboexotica festival.
benjamincowden.com/corpse-reviver

The Italian-made, $100,000 **MakrShakr**
robotic bartender (below), seen in Royal
Caribbean cruise ships and the Tipsy Robot
bar in Vegas, uses two 6-axis Kuka robot
arms to mix, muddle, and shake thousands
of recipes from 158 ceiling-mounted
bottles. makrshakr.com

The **Bartendro 15** drinkbot, Kickstarted and
showcased at Maker Faire Bay Area 2013,
retails for $3,700. We use its open source
software in this Hello Drinkbot project.
(Thanks, Party Robotics!) bartendro.com

Sonos for All

BEN HOBBY
is an 18-year-old maker (and destroyer) from Maryland. When he isn't building 3D printers, CNC machines, and electric scooters, or raiding yard sales, he studies mechanical engineering and music at the University of Maryland, Baltimore County. Find him on Twitter @benhobby.

Written and photographed by Ben Hobby

Hack the Sonos-Ikea Symfonisk to make high-quality networked bookshelf speakers on a budget

TIME REQUIRED:
1–2 Hours

DIFFICULTY:
Moderate

COST:
$100–$150 per speaker, depending on how much you splurge for the recipient speaker.

MATERIALS
» **Ikea Symfonisk bookshelf speaker**
» **Large, efficient, passive speaker** Be sure to select a pair of two-way speakers that are either set up for bi-amping (e.g. four input terminals instead of the standard two), or that you aren't opposed to dissecting.
» **Enclosure** for the electronics (or provisions for mounting them inside the speaker)

TOOLS
» **Screwdriver, Phillips, #2** Also a #1 is useful if you're picky.
» **Pliers, needlenose**
» **Spudger or pick** if you don't have fingernails long enough

CAUTION: When the amplifier is removed, the mains voltage is exposed. Be careful handling it.

(A)

As soon as I saw Ikea's announcement for a $99 Sonos-powered Airplay speaker, I was excited for the possibilities its internals could bring. And oh boy, was I right to be. In this tutorial, I'll show you how to whittle away the woefully inefficient compact speaker unit, and bring Sonos Trueplay to life on a pair of quality vintage bookshelf speakers.

OPEN IT UP

First, make sure the Ikea speaker is unplugged from power. While there isn't much information out there on disassembly of this speaker, a bit of hunting around the sealed plastic enclosure revealed that the fabric tab with the Sonos and Ikea logos functions as a pull tab to get the grille off (Figure Ⓐ).

This reveals the three acoustic elements: a 1" rubber dome tweeter, a 3" woofer, and a curved tuned port. The rubber inserts that suspend the grille in place are firmly seated in their holes, requiring needlenose pliers to remove (Figure Ⓑ). Underneath each insert is a Loctite-sealed #2

(B)

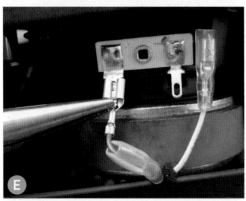

Phillips screw, which will give out a small pop when unscrewed, due to the threadlock releasing.

Next is the tricky part. The front panel, with all the speakers, is sealed to the case by a foam strip that is friction-fit. The easiest way to remove it is to wedge as many fingers as you can into the bass port, hook them around the bend toward the tweeter, and pull firmly and consistently (Figure C). Be careful not to yank the whole assembly too hard, as there are a few connections to the motherboard inside that can be quite fragile.

Once it's free from the foam seal, open it up like a book in order to gain access to the four spade connectors, two per driver, and the small flex cable for the buttons and status lights (Figure D).

DISCONNECT THE ELECTRONICS

There are two different size spade connectors per speaker, a smaller one for ground, and a larger one for the positive terminal. Be careful removing these; they're locking spade connectors, meaning that if you pull directly on them, something important will give instead of the connector. Pull back the clear vinyl insulation and locate the small locking tab. Squeeze this down with the needlenose pliers as you pull, and they'll come off with little to no effort (Figure E).

Next is the tiny, fragile 8-pin flex cable, and corresponding connector, which the motherboard uses to connect to the front panel controls and indicator light. After peeling back the small oval-shaped foam tape covering the connector, use a spudger, or a fingernail, to open the latch, and gently lift out the end of the cable (Figure F).

Now that the front panel is completely free, we get our first real look at the internals of the speaker, and in a shocking twist, we're greeted with the same motherboard as the Sonos Play:1, a speaker that costs twice as much. We'll take a closer look at the motherboard later, but just as a precursor, we can see three large, name-brand capacitors, a socketed Wi-Fi card (meaning a feasible Wi-Fi upgrade years on), and properly isolated high- and low-voltage sections of the PCB (Figure G).

Next you'll remove the brace that holds the foam-wrapped speaker cables in place, a nice touch that prevents the cables from rattling against the enclosure. Two short Phillips screws

G

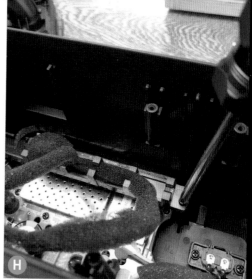

H

hold the brace in (Figure ), and once the brace is free, you can remove the cables and ferrite core from the clips.

The speaker cables have their own socket and connector, but watch out, as it is yet another locking connector. Use the end of your screwdriver to push the small rectangular latch toward the body of the connector (Figure **I**), while simultaneously giving the speaker cable a gentle tug.

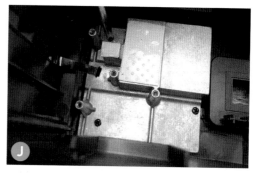

I

On the opposite corner of the motherboard, the AC power input has a similar connector, which is removed in the same way. Next to that are the two Wi-Fi antenna cables, one red and one yellow. It's easiest to remove these by removing the blob of hot glue they're seated in at the same time. If it all comes out in one unit, reassembly outside of the case will be easier. If not, peel away the excess hot glue, and prepare for the microscopic job of reconnecting the two coaxial connectors.

J

Next, remove the six screws holding the motherboard in place. Be careful with the four screws closest to the bottom of the unit, as they are threaded into a cast aluminum heatsink, and require just a bit more torque. Remove the board from the enclosure, turning it diagonally to avoid the internal ribbing.

This beefy heatsink (Figure **J**) is held in by three more Phillips screws, but be careful not to damage the two thermal pads when removing the screws. Remove the heatsink, and use four of the six motherboard screws to affix it back to its spot on the board (Figure **K**).

Now that the motherboard and heatsink are one

K

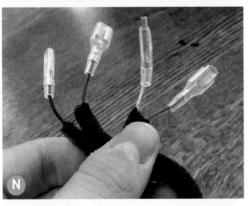

again, we can look closer at the board, and marvel at the quality of it: a good mixture of cheaper and more compact SMD components, as well as appropriately beefy through-hole components. Each side of the board has its own EMI shielded enclosure, with the computer on the front and the amplifier on the back. Under the shielding rest two TPA3116 amplifier ICs, each capable of driving 100 watts at 8 ohms.

Plated mounting holes, heavy doses of potting for insulation, and name-brand components all jump out as features of much more expensive circuitry.

Back to the front panel. You'll find four small Phillips screws holding the button assembly to the panel. Remove these (Figure **L**), and the buttons, rubber membrane, and small daughterboard are freed.

Now, slowly peel the flex cable from the bottom of the enclosure, as it's needed to connect the two circuit boards. It's worth noting that the 8 contact pads of the flex cable face *toward* the circuit board upon reinsertion. Remove the two Phillips screws around the AC socket, and reconnect it to the motherboard for testing.

Next up are the two oddly angled antennas in the top of the enclosure. In my opinion, these feel like an afterthought, as they each have their own plastic bracket adhering them to the enclosure, instead of having their mounts molded into the plastic, like all the other components. Speculation aside, they're easy enough to remove; using a bit of force on the bottom of each bracket breaks the strong double-sided tape holding them in (Figure **M**), and the hot glue holding the antenna cables into the enclosure is not too hard to break.

BUILD IT ANEW

Now that you have all the electronics free, it's time to connect a speaker and give it a quick test. This is where things start to diverge. As mentioned, you will need either a bi-amp-able speaker (i.e., separate terminals for woofer and tweeter) or a two-way speaker for which you can bypass the internal crossover. The Symfonisk uses two discrete amplifier ICs, one for the tweeter, and one for the woofer. Bridging them not only causes the amplifier to shut down, but it may damage your $99 pile of circuit boards and wires.

The connections on the Symfonisk board are as follows: for the tweeter, red for positive, black for negative; and for the woofer, blue for positive, white for negative (Figure ). A polarity swap isn't the end of the world, it'll just cause your two-way speaker to be out of phase with itself. I have my board connected to a vintage JMLab Chorus 707 bi-amped two-way bookshelf speaker, as shown in Figure O.

> **CAUTION:** Before plugging into AC power, take care to avoid contact with any high-voltage connections or components of the amplifier unit.

Plug in the AC cord, staying clear of the live connections on the circuit board, and launch the Sonos app. Begin setting it up as a new speaker, following the directions until you get to the Trueplay tuning. Follow the prompts, disregarding that your speaker isn't in its final resting place, and allow the speaker to chirp and buzz while you wave your phone around the room like the app shows. You can easily retune the speakers once the setup is finalized. The Trueplay algorithm quickly adapts the digital crossover within the board, allocating the correct frequency ranges to the woofer and the tweeter of your speaker, making it sound near perfect.

And there you have it! One half of the perfect budget Sonos system. Repeat the procedure for your second speaker. Build yourself a nice enclosure for the electronics or tuck them inside with strategic placement for the controls and jacks, and revel in the glory of a beautiful stereo pair of Airplay 2, Sonos-enabled, algorithmically tuned speakers of your choosing for a few hundred bucks. ●

Finished Touches

Following Ben's how-to at makezine.com, Vladimir Tombette Sonos-ified his Focal Chorus 706S speakers and shared his notes:

The two class D amplifiers are TPA3116 (100W @ 8Ω each).

Minor build challenges:
- You need to extend the internal AC and speaker wires
- Requires routing of wood and plastic
- Sealing the speaker again is tricky.

Overall, it sounds great. Trueplay is a miracle.

Flat-Out Flinger

Written and photographed by Jonathan Stapleton

JONATHAN STAPLETON has taught middle and high school science for 22 years, and was briefly a toy designer with his invention, Reptangles. He enjoys reading, inventing, fishing, hunting, disc sports, music, and swimming in the lakes and rivers near Essex, Vermont.

Build the world's newest, simplest trebuchet and break some chuckin' records!

This is the world's newest trebuchet design and it's already setting new hurling records. Not only is the Walking Arm Trebuchet simple, it's also extremely efficient. Using this trebuchet, my 8-year-old son set the all-time record for best design at the 2018 Vermont Pumpkin Chuckin' Festival. His lightweight 20-pound, 41-inch tall trebuchet threw a 3-ounce ball 266 feet — scaled up, that's equivalent to 780 feet in the heavyweight division. My 500-pound, 10-foot version threw a 5-pound cantaloupe over 700 feet — also a new record — but it went so high and so far that the spotter never saw it pass over him on its way into the woods!

This year my 500-pounder threw a 5-pound rice-filled soccer ball 870 feet (and it rolled to 975) but unfortunately at the 2019 Pumpkin Chuckin' it threw so hard that all the real fruits were crushed during launch. My son won the grand prize again.

The instructions provided here show how to make a 20-pound, 41-inch version, like my son's. The dimensions were chosen to adhere to the lightweight division rules at the Vermont festival. As you'll see, my son did all the work himself. I did help out by holding some things in place and showing him how to complete the steps.

I've made other versions of this trebuchet with longer arms, mostly for throwing the excess apples that fall in our yard. While they would not be legal at our state competition, they can really whip an apple.

TIME REQUIRED:
A Weekend

DIFFICULTY:
Easy

COST:
$40–$60

MATERIALS

WOOD:
- » Pine board, ½"×3"×36"
- » Lumber, 2×4, 12" length
- » Square dowels, ¾"×36" (3)
- » Popsicle sticks (2) plus extras

HARDWARE:
- » Lag screw, ½"×6"
- » Heavy washer, ½"
- » Nails, 3" (3)
- » Drywall screws: 2" long (2) and 3" long (8)
- » Hex bolts, ¼":1" long (1) and 2½" long (1)
- » Washers, ¼" (4)
- » Hex nut, ¼"
- » Wing nut, ¼"

MISCELLANEOUS:
- » Thin cord or twine, at least 10' We used duck decoy anchor cord.
- » Cable ties (2)
- » Fabric, 10" square We used ripstop nylon.
- » Iron weights, 17½lbs total We used one 10lb, one 5lb, and one 2½lb. Any weights can be used, but they will probably require modifications to the design. My daughter made a counterweight with concrete from the hardware store. Generally, denser weights are best, so iron is preferable to concrete, sand, or rock.
- » PVC pipe, ¾", 3" length (optional)

TOOLS
- » Measuring tape and pencil
- » Saw
- » Hammer
- » Power drill/driver
- » Drill bits, assorted
- » Driver bit, Phillips head
- » Wrench, ¾" (19mm)
- » Scissors
- » Hacksaw or bolt cutters to cut the head off a nail

OPTIONAL:
- » Center punch
- » Large file
- » Torch and locking pliers for heating and holding a hot nail
- » Slow motion video camera for fine-tuning trebuchet performance
- » Sharpie marker for drawing on fabric

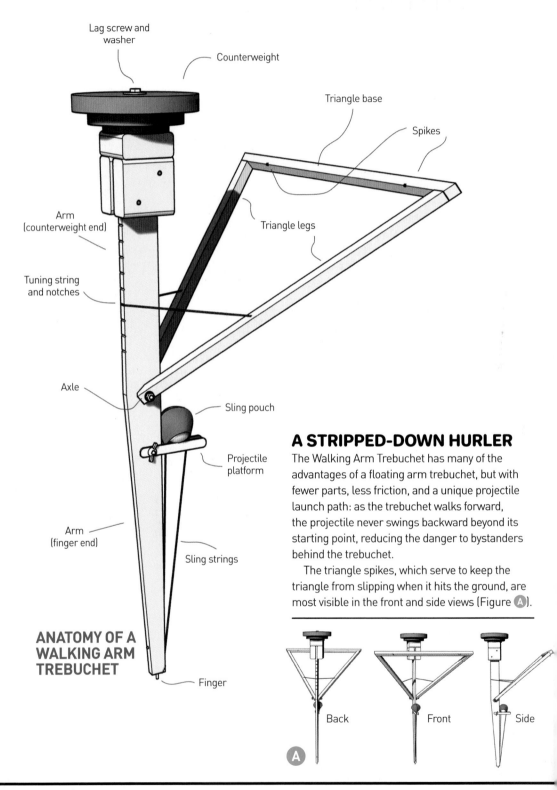

Lag screw and washer

Counterweight

Triangle base

Spikes

Arm (counterweight end)

Triangle legs

Tuning string and notches

Axle

Sling pouch

Projectile platform

Arm (finger end)

Sling strings

ANATOMY OF A WALKING ARM TREBUCHET

Finger

A STRIPPED-DOWN HURLER

The Walking Arm Trebuchet has many of the advantages of a floating arm trebuchet, but with fewer parts, less friction, and a unique projectile launch path: as the trebuchet walks forward, the projectile never swings backward beyond its starting point, reducing the danger to bystanders behind the trebuchet.

The triangle spikes, which serve to keep the triangle from slipping when it hits the ground, are most visible in the front and side views (Figure Ⓐ).

Back

Front

Side

Ⓐ

EVOLUTION OF THE WALKING ARM

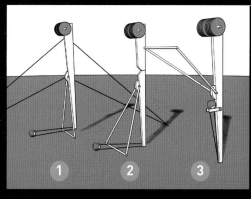

1   2   3

The Walking Arm Trebuchet's evolution was driven by the unique rules of the Vermont Pumpkin Chuckin' Festival. Dave Jordan, founder of the festival, set a limit on *overall trebuchet weight*, rather than the counterweight. If it hadn't been trying to keep my design as light as possible, I never would have invented this trebuchet.

I wanted a design that would allow the counterweight to fall as far as possible, to maximize potential energy, with a long throwing arm and minimal friction. I also wanted the trebuchet to "shift gears" on its own.

DESIGN #1

The *floating arm trebuchet (FAT)* invented by Ron Toms seemed like a good place to start, because the FAT's weight falls nearly its entire height and it "shifts gears": because the throwing arm's axle *floats*, rather than having a fixed axis, each small movement of the falling counterweight results in an increasingly greater movement of the projectile. (Adding wheels to a medieval *hanging counterweight trebuchet (HCW)* has a similar effect.)

My first attempt didn't walk. It was an arm standing on end, with the weight on top, supported at its center by an axle. The axle was on a simple stand that pivoted on the ground, allowing the

axle itself to swing forward or backward. As the counterweight's inertia carried it downward in a relatively straight path, the axle and arm were forced backward, then pulled forward again. This motion flung the projectile in a path similar to that of a FAT, but without all the extra structural weight.

The main problem with this design was that it required high-tension guy wires to hold everything in the right launch position. And creating a trigger to release all the parts simultaneously was tricky.

DESIGN #2

I added a joint in the arm, like the *King Arthur (KA) trebuchet* invented by Chris Gerow. This let me get rid of the guy wires. At rest, the trebuchet had become a tripod, with the throwing arm standing on the ground behind an axle stand that extended forward. When fired, the forward inertia of the arm's top segment pulled the whole trebuchet with it, causing the arm to walk forward and swing over the axle. I was smitten by the simplicity and the walking nature of this design. My daughter used it to win the lightweight division in 2015, but it just didn't throw things as far as the first design. It was also harder to tune. So I tried to marry the two.

DESIGN #3: EUREKA!

Eventually I realized that if I lifted the axle stand off the ground, the trebuchet would tip forward onto the axle stand, walk, and throw. It worked, but not until I added spikes to keep the axle stand from slipping on the ground. The axle stand that had been rooted to the ground in Design #1 became the triangle in my fully evolved Walking Arm Trebuchet.

The idea of staging the projectile on a platform attached to the arm came from a *whipper* trebuchet, a HCW variant invented by the late, great Raymond "Ripcord" Goodsell, that I'd seen — where else? — at the Vermont Pumpkin Chuckin' Festival.

Hanging counterweight trebuchet, floating arm trebuchet (FAT), King Arthur trebuchet (KA), and whipper trebuchet

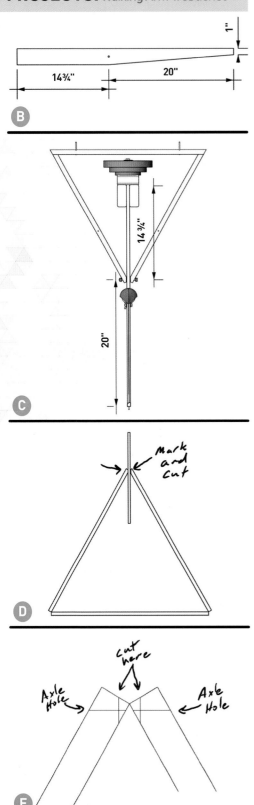

1. MAKE THE THROWING ARM

Make the arm from the ½"×3"×36" pine board. Drill a ¼" hole centered 20" from one end, as shown in Figure Ⓑ. This will be the finger end of the arm. Taper the arm from a full 3" at the axle to 1" at the finger end. We used a band saw to cut the board, but a handsaw would work.

Now determine the length of the counterweight end of the arm. Figure B shows the counterweight end extending 14¾" from the axle hole. That was the length that worked for us. That distance may need to be different for your arm, depending on the height of your stack of weights, because the weights need to clear the triangle, as you can see in Figure Ⓒ. To determine how long to make the counterweight end of your throwing arm, measure the height of your stack of weights and subtract that number from 17¼". That should give you nearly 1" of clearance for the counterweight to swing through the triangle. If you allow too much clearance, then your weight doesn't fall as far, so you're giving up energy.

Using the length you just calculated, measure from the center of the axle hole toward the non-tapered end of the board. Make a mark and cut off the board.

2. MARK AND CUT THE TRIANGLE

The point here is to build a strong triangle with a base 20½" inches from the axle. The triangle needs to be wide enough so that the counterweight can safely swing through without hitting it, and it needs to be strong enough to withstand extreme forces during the launch. To keep it simple, we made an equilateral triangle with plenty of extra room; feel free to make your triangle a little narrower.

Lay the three 36" square dowels on a floor and arrange them in a triangle. Lay a ½" scrap on top of the triangle legs as shown, and use it to mark the locations where you'll cut (Figure Ⓓ). It's important that this scrap is ½" thick because it represents the thickness of the throwing arm.

Now mark the axle hole (Figure Ⓔ). This mark should be perpendicular to the cut marks that you already made, and, therefore, perpendicular to the throwing arm.

Without moving the legs of the triangle, shift the triangle base toward the axle mark until the bottom of the base is 20½" from the axle mark (Figure **F**). Make sure you still have an equilateral triangle; the legs should be the same length. Our legs ended up being about 23⅜" inches, when measured as shown.

Keeping the legs in this position, mark both legs and the base for cutting (Figure **G**). Then cut off both ends of the triangle legs, and both ends of the base.

3. DRILL AND ASSEMBLE THE TRIANGLE

Drill ¼" axle holes, centered in each leg, parallel to the lines you drew. It's easiest to drill the side where the hole will be perpendicular to the wood. To be extra sure he drilled in the right spot, my son first carefully made a pencil mark where he wanted to drill. Next he made an indentation with a center punch, and then he pre-drilled with a ⅛" bit before finally drilling with the ¼" bit (Figure **H**).

Hot-glue the legs to the base as shown in Figure **I**, then secure these joints with drywall screws. First drill ⅛" pilot holes, then screw 2" drywall screws into the pilot holes. You're drilling near the ends of the dowels, so you may split the wood. We split one end, then glued and clamped it. We've had no problems with it.

4. BUILD THE PROJECTILE PLATFORM

Drill a ¼" hole 4" from the axle hole, on the finger side of the arm, as shown in Figure **J**. Our hole was about ¾" from the edge of the arm.

Drill a ¼" hole centered near the end of a popsicle stick. To reduce the chance of splitting, clamp the popsicle stick firmly against a piece of waste wood, and drill through the stick and into the wood. You may want to drill a small pilot hole first, to make sure your hole is in the center of the stick. (Popsicle sticks are handy but they're barely big enough to drill a ¼" hole. Any thin, strong, drillable material would work well here. It just needs to hold up the projectile.) Repeat this with one more popsicle stick. While you're at it, make some extras.

Use the 1" hex bolt and the wing nut to attach the popsicle sticks to the arm (Figure **K**).

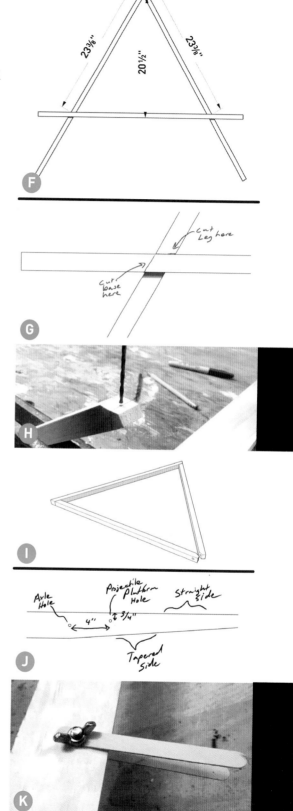

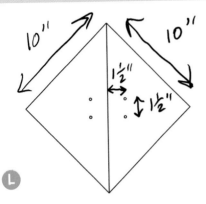

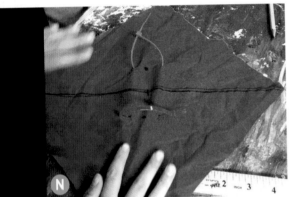

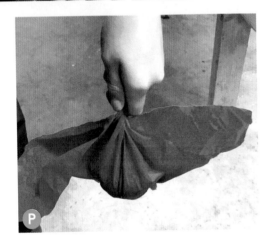

5. MAKE THE SLING POUCH

We used fabric for our design because it is readily available. For my son's competition trebuchet we used a piece of net, which seems lighter and has less wind resistance.

Cut a 10"×10" square of fabric. Nylon works well because you can melt the holes in it and seal frayed edges. Draw a diagonal line from one corner to the opposite corner. Find and mark the center of the diagonal line.

Draw two pairs of dots. Each pair should be about 1½" apart, and 1½" from the diagonal line (Figure **L**).

Now make holes where you drew the dots. Because we used nylon, we could melt holes with a nail that we heated with a propane torch. My son held the nail firmly with a pair of vise-grip pliers (Figure **M**).

At each end of the diagonal line, tie the corner of the fabric in a knot, tight and close to the edge of the fabric. Then use cable ties to cinch each pair of holes together (Figures **N** and **O**). Clip off the cable tie excess.

Test your pouch. Place your projectile of choice

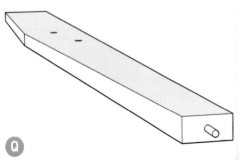

in the pouch (Figure P). We used a "high bounce Pinky" ball for the projectile. A small apple or just about any other roundish object should work. See if the projectile stays in place when you hold both ends of the pouch. Then let go of one end of the pouch and see if the projectile rolls out. If the projectile either won't stay in or won't come out, cut the cable ties and make new pairs of holes. Spacing the holes more widely (before cinching with cable ties) should hold the projectile more securely in place. More closely spaced holes should help it roll out. Placing the holes closer to the centerline may also help the projectile roll out more easily.

Trim off excess fabric.

6. MAKE THE FINGER

Cut the head off one of the nails, so that the remaining headless nail is about 2" long. This will be the trebuchet finger (Figure Q). It looks like a round dowel in this drawing, but for this small-sized trebuchet we only need a nail.

> **TIP:** If you can't cut the nail, you could try using a finishing nail. The minimal head on a finishing nail could be sanded off, or could even be left as it is, without causing much trouble with the sling release.

Drill a small pilot hole into the finger end of the arm (Figure R). The pilot hole should be a little narrower than the nail that will become the finger. The purpose of the pilot hole is to prevent splitting as you hammer in the nail. It would be a good idea to try a practice hole on some waste wood first.

Nail the pointy end of the clipped nail into the pilot hole, so that ½" sticks out. If the end that's sticking out is sharp, file it down until it is rounded. You don't want it to scratch someone.

7. TIE THE SLING CORDS

Tie a long length (maybe 30") of cord next to one of the sling knots (Figure S). The knot in the sling fabric keeps the knot in the cord from slipping off.

In the same way, attach another length of cord to the opposite corner.

8. ATTACH THE SLING TO THE ARM

Drill a hole through the throwing arm, barely larger than the cord, centered 2" from the finger end of the arm (Figure T). This hole needs to be

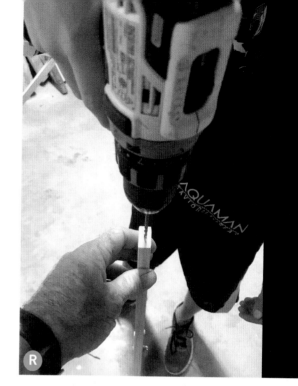
R

S

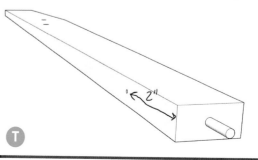
T

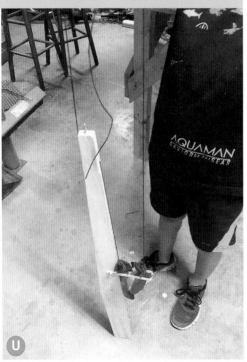

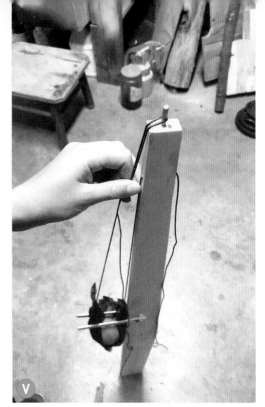

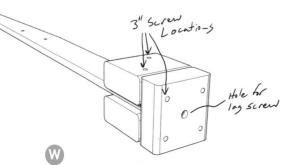

far enough from the end so that it doesn't hit the buried end of the nail that forms the finger.

Insert one sling cord through the hole so that it enters the arm on the flat side and comes out on the tapered side. Place your projectile in the sling pouch, and arrange both cords as shown in Figure U. Snug the projectile into its position beneath the platform (popsicle sticks) and make the cords taut. The popsicle sticks should stick out perpendicular to the arm.

Mark the point where the sling cord exits the tapered side of the arm, and then tie a knot about ½" beyond that point. The purpose of this knot is to prevent the cord from pulling through the arm. After you make the knot, check to see that the sling cord is the correct length. If it's too long, shorten it by tying one or more additional knots near the first one.

Take the other sling cord, loop it around the finger (Figure V), and tie a knot to create a loop at the end of the sling cord.

At this point, you have created a weapon called a "staff sling." A video on the Instructables page for this project shows my son testing his, to make sure the projectile platform and sling work properly. Don't overdo it on the first try. My daughter tried it, and our Pinky ball is now lost in the woods.

9. BUILD THE COUNTERWEIGHT BASE

The counterweights need to have a strong base connecting them to the arm (Figure **W**). To make it, cut three 3½" lengths of 2×4 lumber. Sandwich two of the 2×4s around the end of the arm, and screw the sandwich together with 3" drywall screws (Figure **X**). Don't put a screw in the center, because it may interfere with the lag screw that will hold the weights in place.

> **TIP:** These short 2×4 pieces will split easily. To prevent splitting, drill pilot holes for the screws.

Attach the third 2×4 to the ends of first two. Don't put a screw in the center of this one either, because that's the location of the lag screw. You've built your counterweight base.

Now drill a hole into the center of the counterweight base, a little bit smaller than the lag screw (Figure **Y**). I think we used a ⁵⁄₁₆" bit for the ½" lag screw, but a ⅜" should also work.

You may want to drill a test hole in some scrap wood first, and try the screw's fit. You'll want it to be nice and tight.

10. ATTACH THE WEIGHTS

Stand up the arm on its finger end. Stack the weights with the smallest (2½lb) on the bottom. The 5lb goes next and the largest (10lb) goes on the top (Figure **Z**). Why? You want the largest weight to fall as far as possible, to give your projectile the most energy.

> **IMPORTANT:** Stand your trebuchet on grass or dirt. If you do it on concrete, you may mash the finger in too far. You can also drill a hole in a piece of wood and stand the arm on the wood, with the finger in the hole.

I cut a section of ¾" PVC pipe and placed it in the weights' holes. That helped keep the weights aligned, but it isn't necessary.

Place the ½" washer on the 6" lag screw. Tighten the screw to clamp the weights in place by screwing it tightly into the counterweight base (Figure **Aa**). The screw should take a ¾" or 19mm wrench.

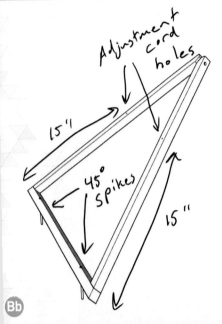

Adjustment cord holes

15"

45° Spikes

15"

Bb

Cc

Dd

Ee

11. ADD THE TRIANGLE SPIKES

The spikes run 45° diagonally through the triangle base, corner edge to corner edge, as shown in Figure **Bb**. Make these holes about 6" from the triangle legs (far enough so the legs won't interfere with your drill or hammer).

To make drilling into the edge easier, first sand, cut, or file a notch into the triangle base where you want to drill (Figure **Cc**). My son also likes to use a center punch to make an indentation before drilling (Figure **Dd**). Drill holes small enough to keep the spikes tightly in place, but big enough to prevent splitting; we drilled ⅛" holes (Figure **Ee**).

Hammer the remaining two nails through the holes. File or cut the ends of the spikes so they're not so sharp and scratchy. They just need to be able to stick into grass or dirt.

> **NOTE:** We like the spikes because they keep this trebuchet simple, but they're not absolutely necessary. My 500-pound trebuchet doesn't have spikes; it falls onto an anchored board with a lip that prevents the triangle from sliding forward (see "Scale It Up" on page 81). We have also arrested the triangle by running cords from the triangle to a launch platform that's anchored in the ground (see "Competition Launch Setup," page 80).

12. ADD THE TUNING STRING

Mark each leg 15" from the base corner, as shown in Figure Bb. At each mark, drill a hole barely larger than the cord.

Cut a length of cord at least 24". Tie a knot in one end, insert the other end through both leg holes, then tie a knot in the free end so it won't pull back through, so that the length of cord between the two legs is 16" (Figure **Ff**).

> **NOTE:** The exact dimensions described in this step aren't entirely important. They worked for our trebuchet, but other dimensions would also work. If you understand the purpose of this string, you can probably disregard our numbers.

13. ATTACH TRIANGLE TO ARM

Attach the triangle to the arm using the 2½" hex bolt, as shown in Figure **Gg**. Use a washer on each side of each piece of wood — 4 washers in all. Turn the nut by hand until it begins to tighten; don't

Ff

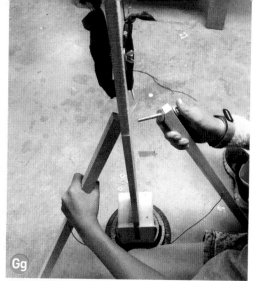

Gg

Hh

overtighten it. At this point, the total weight of the trebuchet should be around 20 pounds.

> **IMPORTANT:** The axle joint should be loose because the triangle doesn't always hit the ground at the angle you expect. The counterweight, on the other hand, has so much inertia that it follows a fairly predictable path. Keeping the joint loose allows the counterweight to pass through the triangle in a wide range of orientations. If the joint were tight, their relative positions and motions would be restricted and the counterweight would be more likely to break the triangle.

14. CUT THE TUNING NOTCHES

Using a file or a saw, make a series of notches on the back of the arm (Figure **Hh**). These notches are used to hold the tuning string in place. To begin, position the string so that the triangle makes a 45° angle with the arm, and the first notch here. Make other notches on other sides of this notch. It's helpful to label the notches (Figure **Ii**) so that you can keep track of where you set the tuning string during each launch.

Your Walking Arm Trebuchet is complete!

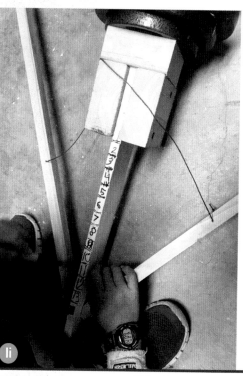

Ii

AT MY SIGNAL, UNLEASH FUN!

LOAD AND FIRE

To load your trebuchet, stand it upside down — with the counterweight on the ground. Put a tennis ball in the sling pouch, position the pouch below the popsicle sticks, and loop the sling over the finger. Adjust the sticks to hold the pouch snugly.

On grass or dirt, turn the trebuchet right-side up (counterweight on top) and hold it there while you set the triangle string into a tuning notch of your choice. Aim your trebuchet downrange and make sure it's all clear.

To fire, just nudge the trebuchet forward so that it falls squarely onto the triangle ... step back, and watch it walk and hurl!

TUNE IT UP

There are a few relatively easy ways to fine-tune the trebuchet's performance. Here they are in order of decreasing ease and utility.

- **Adjust the triangle's position before firing** by moving the tuning string to different notches. If you have a camera with slow-motion capability, take videos of your launches. As a general rule, the counterweight should be at its lowest point when the projectile releases. The counterweight should also be moving as slowly as possible at this point.

  If the trebuchet flops over forward after firing, or if the shot has a very low trajectory, the triangle is probably too low. If the projectile is released early, flies too high, or the counterweight hits the ground before the projectile is released, the triangle is probably too high.

- **Adjust the sling length**. This is more of a hassle. It requires not only changing the sling, but also moving the projectile platform.

- **Adjust the finger**. I haven't needed to adjust the fingers on any recent trebuchets of this type, but bending the finger forward or backward is one way to control when the projectile is released. Bending the finger forward holds the projectile in its sling longer. Bending it backward causes the projectile to be released earlier.

COMPETITION LAUNCH SETUP

In competition we fire this trebuchet from a platform (Figure **Jj**), and we release it with

a simple "trigger" mechanism (Figure Ⓚ). Without the platform, the trebuchet doesn't fall from as high as it is legally entitled to fall. The maximum allowed height is 41", but the top of the counterweight is probably at about 39" when we just stand the trebuchet in the grass. The triggering system is in place to prevent us from adding any extra energy by giving the trebuchet a push.

The racquetball that my son launched in competition was partially filled with water to bring it up to a regulation weight of 3 ounces. The water was inserted using a needle and syringe. Water-filled racquetballs make fun projectiles.

SCALE IT DOWN

In response to one of my Instructables readers, I made a small 3D-printed version of the Walking Arm Trebuchet (Figure Ⓛ). I was going for a desktop version, but really it seems at home on carpet. It's a rough version 1.0, but I posted the *.stl* file on Instructables if you'd like to print one.

I used pieces of a large paper clip for the spikes and the finger. For good measure, I sharpened the spikes with sandpaper. I used fishing line for the sling and the tuning string. There's no pouch; the sling is just a loop of fishing line that's permanently attached to the projectile. They fly off together. The proportions are different from the larger versions, which may be why I needed to bend the finger forward on this one to get a good release angle.

SCALE IT UP!

My heavyweight competition trebuchet is 10' tall and 500 pounds (Figure Ⓜ). The 5-pound cantaloupe that it threw was never found! On its debut two years ago, the big trebuchet was so heavy that it sank into the ground and broke itself — repeatedly. So last year I modified it: I set it up so that the triangle lands on a 1½" thick plywood pad and is stopped by a lip made of 2×6s. ◉

You can read more about these trebuchets' success on the Vermont Pumpkin Chuckin' site at vtpumpkinchuckin.blogspot.com/2018 and /2019, and see them in action at instructables.com/id/Worlds-Simplest-and-Newest-Trebuchet, where you'll also find more assembly diagrams, photos, and videos.

Ⓛ

Ⓜ

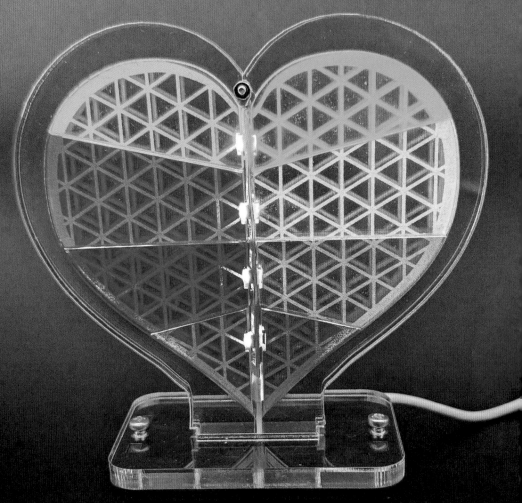

"Inner Glow" LED Heart

Create brilliant displays and décor with a new edge-lighting trick

Written and photographed by Debra Ansell

DEBRA ANSELL (geekmomproects.com) believes LEDs improve everything. Her newest and most challenging project is a commercial line of codeable wearable LED bags and backpacks.

Edge-lit displays typically consist of an etched sheet of acrylic, with one edge embedded in an opaque base containing a light source. The light diffuses through the acrylic to make the etchings glow. Different colors applied around the edge will blend smoothly inside; you can see this effect in my Edge-Lit LED Rainbow project in *Make:* Volume 69 (makezine.com/go/edge-lit-led-rainbow).

My new project takes a different approach to edge lighting, by cutting the acrylic into jigsaw-puzzle-like sections and embedding the light source between the pieces. This technique makes the shape appear to glow from the inside out. Illuminating each piece with its own color creates sharp boundaries for an unusual effect that's great for dynamic, colorful animations. If you do your own laser cutting you can make it for 30 bucks.

Here I'll show how to build an inner-glow heart, in plenty of time for February gift giving (hint hint). I also provide two other designs, and you can easily customize this project to create your own.

1. BUILD THE WOOD BASE
Laser-cut the base pieces from ⅛" wood (Figure Ⓐ), using the template in *WoodPieces.svg*. If you're using veneered wood, you can use a single-sided veneer as only one side will be visible.

Spread a little wood glue on the bottom edges of all side pieces and slot them into the bottom piece (Figure Ⓑ). Don't glue the top, as it needs to be removable to place the microcontroller inside.

2. LASER-CUT THE ACRYLIC
There are three layers of acrylic. The 1/16" outer layers (*HeartSixteenthInchAcrylic.svg*) serve to hold

TIME REQUIRED:
1–2 Hours

DIFFICULTY:
Intermediate

COST:
$30–$60

MATERIALS
» **Cast acrylic sheet, clear, ⅛" thick, 6"×8" or bigger** Cast acrylic glows better than extruded acrylic.
» **Cast acrylic sheet, clear, 1/16" thick, 6"×10" or bigger**
» **Wood or plywood, ⅛" thick, at least 6"×6"** for the base
» **Mini RGB LED pixels (8) 4mm–6mm wide** (less than ¼"), cut from a "skinny" RGB LED strip, WS2812B or SK6812 3535 type, 60 LEDs per meter. I used Adafruit NeoPixel strip #2959 with the casing removed, making it 5mm wide.
» **Machine screw, M2×10mm, with nut**
» **Machine screws, M3×20mm, with nuts (2)**
» **Hookup wire, 26AWG, white**
» **Microcontroller** to control the RGB pixels. I used an Adafruit Trinket M0.
» **Micro USB cable**
» **Copper or mylar tape, 3/16" wide**
» **Wood glue**
» **Heat-shrink tubing** a small piece

TOOLS
» **Laser cutter (optional)** or use an online cutting service like Ponoko. Download the free cutting files at makezine.com/go/inner-glow-heart.
» **Wire cutter and stripper**
» **Soldering iron and solder**
» **Heat gun for the shrink tube**
» **Sharpie markers in three colors**
» **Screwdriver**

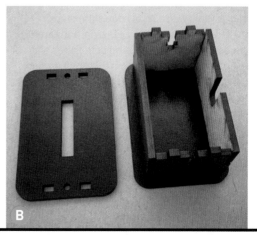

Ⓐ

Ⓑ

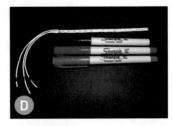

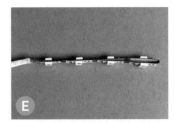

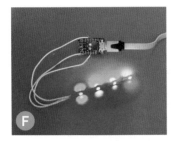

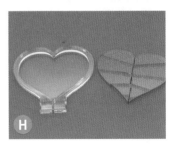

in place the ¼" pieces in the inner layer, *HeartQuarterInchAcrylic.svg* (Figure **C**). There's also a ¼" acrylic base to support this assembly.

Cut the shapes in both files. When laser cutting the ¼" acrylic, *etch* the pink shapes and *cut* all the black outlines. Several different etching patterns are included to choose from — or create your own. Remove any protective paper from the acrylic.

3. PREPARE THE LED STRIP

Cut a length of 8 LEDs from the LED strip. Cut three 4" lengths of 26 AWG white hookup wire, strip one end, and solder them to the three pads at the data input end of the LED strip. The white wire almost vanishes within the clear acrylic, which is aesthetically pleasing, but it can be hard to distinguish between power, ground, and signal. I use different color Sharpies to mark the wires (Figure **D**). To provide strain relief, cover the solder joints with a bit of heat-shrink tubing.

Find the middle of the LED strip, and carefully bend it in half so the LEDs on both halves sit back to back. When the alignment is correct, remove the adhesive backing from the strip and stick the two halves together (Figure **E**).

> **IMPORTANT:** Don't crease the strip at the bend. It's OK if there's a small loop at the end.

Strip the unsoldered ends of the hookup wires and connect them to your microcontroller to be sure the LED strip lights up. With the Trinket

M0 it's easy just to slip the wires into their corresponding pins — LED Power to Trinket 3V, LED GND to Trinket GND, and LED Signal to Trinket Pin D1 — and hold them in place for a few moments to verify the connections (Figure **F**).

4. MASK THE INNER ACRYLIC PIECES

Because the inner acrylic pieces are small, light may leak between them. To create distinct edges, you'll mask off their outer edges with reflective tape. You can use ³⁄₁₆" copper or mylar tape.

Take each of the etched pieces and cover every side with reflective tape except the side containing the notch for the LED (Figure **G**).

5. ASSEMBLE THE ACRYLIC LAYERS

Place one of the ¹⁄₁₆" outer layers flat on a tabletop, with a 10mm M2 screw protruding vertically through the small hole near the top.

Next, assemble the middle layer. Start with the large ¼" outline, aligning it atop the outer layer so the screw passes through the hole (Figure **H**). Assemble the other ¼" pieces inside the outline, making sure their etched sides all face the same way. Nudge the pieces away from the center to create a gap where the LED strip will go.

6. EMBED THE LED STRIP

Lay the folded LED strip carefully within the gap so the LEDs fit neatly into their notches (Figure **I**). Then lay the final ¹⁄₁₆" layer of acrylic across the top to hold it all together. Secure the layers

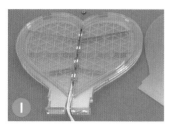

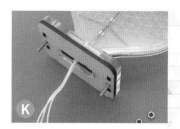

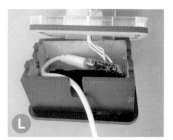

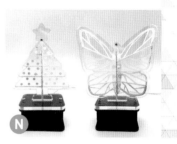

together by fastening the M2 nut to the screw.

Hold the layers of the heart together so the tabs at the bottom align. The middle tab will protrude below the outer tabs. Slide the wires and all three tabs through the rectangular hole in the ¼" acrylic base (Figure **J**). The middle tab has a slot in the center; you'll need to squeeze the edges together make it fit. Don't force it, just squeeze gently and wiggle the base until it slips over the tabs.

Now gently slide the wires and the middle tab through the wooden box lid. Again, you'll have to gently squeeze the middle tab to make it fit. Slide the 20mm M3 screws through the corresponding holes in both layers of the base (Figure **K**).

7. CONNECT THE MICROCONTROLLER

Solder the wires from the LED strip to the microcontroller. I soldered the ends to the Trinket M0, and attached a Micro USB cable to the Trinket for power and programming. Place the microcontroller into the wooden base so that the USB cord extends out the back slot (Figure **L**).

8. CLOSE UP THE CASE

Slip the M3 nuts just over the ends of the 20mm screws, place the lid on the base so the nuts fit into the their slots (Figure **M**), then tighten the screws to secure the heart and lid onto the base.

9. PROGRAM THE MICROCONTROLLER

You're ready to light it up. CircuitPython-enabled microcontrollers like the Trinket M0 make it easy to generate colorful dynamic LED patterns with only a few lines of code. For an easy jump start, I adapted Adafruit's CircuitPython NeoPixel code from their Learning Guide at learn.adafruit.com/circuitpython-essentials/circuitpython-neopixel. Only one line of code requires changing, since the example code is already written for a sequence of eight LEDs. Instead of line 6:

```
pixel_pin = board.A1
```

change the LED signal pin to **board.D1**:

```
pixel_pin = board.D1
```

or whatever pin on your microcontroller provides signal to your LED strip.

THE LIGHT WITHIN!

As soon as you save your code, you should see the acrylic heart light up in a series of solid colors and rainbow patterns. Spend a little time with the CircuitPython NeoPixel learning guide and soon you'll be writing your own patterns to make this heart your own.

CHANGE IT UP

You can easily make different inner-glow edge-lit shapes by adapting the outer contours of the acrylic pieces. I made a Christmas tree whose etched "ornaments" are illuminated in different colors, and a butterfly, just for variety (Figure **N**).

As long as you leave notches for the LED strip in the same place, you can adapt this design to come up with your own original decorations to match the season, for gifts or trophies, or just for fun. ⊘

World's Best Paper Airplane

Written and photographed by John Collins

Fold and fly the Guinness-record glider "Suzanne"

All great planes used to be named after great women. My nod to that tradition was naming my world-record paper airplane design after my wife.

Suzanne represents a number of firsts. It's the first glider to hold the record for distance, thrown by Joe Ayoob in 2012. It's the first time a thrower/

designer team has held the record. (Predictably, the old world record holder hated the idea, but Guinness liked it.) It's also the first paper airplane to use a variable dihedral angle to optimize lift-to-drag over a range of speeds.

There's still a $1,000 reward to use my plane to break the Guinness World Record for distance. We know the plane goes well beyond the current record of 226 feet 10 inches; Joe has thrown the plane more than 240 feet.

We hope you'll build and fly the Suzanne glider. We'd love to know just how far the plane can go.

TIME REQUIRED:
15–30 Minutes

DIFFICULTY:
Easy

COST:
$1

MATERIALS

» **Paper, A4 size or US letter size** For official world record competition, I use the maximum weight (100gsm A4) by Guinness rules. The smoothest, stiffest paper stock I could find was Conqueror Paper CX22 Diamond White, unwatermarked.

   If you don't have A4, remove 19mm from one side of a U.S. letter size sheet of paper so that the paper looks taller. 26lb paper is the closest in thickness and weight.

» **Adhesive tape, 25mm wide** such as Scotch Magic tape

TOOLS

» **Scissors**

» **Hobby knife or chopstick** for transferring tape

» **Folding tool** aka bone folder, or paper creasing tool. My favorite is made from real bone. Plastic works, but gets nicked easily. The bone tool stays smooth and makes great creases for years.

» **Snack bag clip with rubber grip** for holding the plane during adjustments

» **Protractor** to measure dihedral angles

» **Cardboard dihedral tool** Use the protractor to cut a piece of cardboard: one side with a 165° angle and the other side with a 155° angle.

TIP: Use an old shaving kit to hold your plane-making tools.

JOHN COLLINS (aka The Paper Airplane Guy) is an award-winning keynote speaker, science educator, and author of four books of original paper airplane designs. He travels all over the world demonstrating fluid dynamics and fun with his innovative paper planes.

John Collins

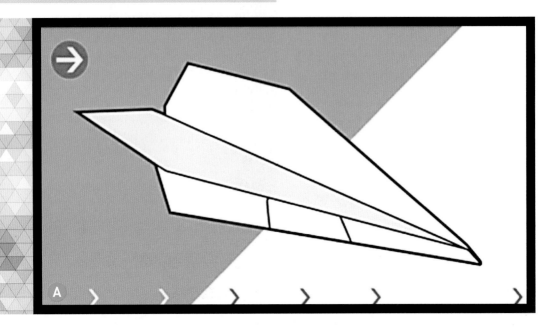

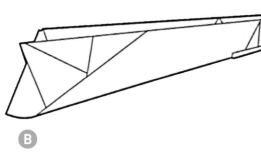

FOLD YOUR WORLD RECORD PAPER PLANE

The folding is easy (throwing for a world record is not), but the taping will take some time.

All folds are "valley" folds (— — — —) except for the "mountain" fold (— •• — ••) in Step 9.

This tutorial is excerpted from my book *The World Record Paper Airplane and International Award-Winning Designs* which you can find at thepaperairplaneguy.com/shop.

GLIDER OR DART?

Previous record holders were very small-winged *darts*. The wings were more like the fins on an arrow; simply for directional stability and not providing lift. It didn't matter if the plane rolled, which they frequently do. The throw was a simple 45° launch, like any ballistic projectile.

For that reason, those kinds of planes were called *ballistic darts* (Figure B). The old record holder, a ballistic dart, took only 3 seconds to go 207' 4".

My plane is a *glider* (Figure A). It takes 9 seconds to fly 226' 10". My plane gets launched level, climbs on its own by generating lift, and really flies the last third of the flight. It gently touches down and skids to a stop. The old kind of plane simply crashes into the finish line.

1. Make a diagonal fold. Line up the top with the left side.

2. Unfold Step 1.

3. Make a diagonal fold the other way.

4. Unfold Step 3.

5. Line up the left edge with the diagonal crease.

6. Line up the right edge with the other diagonal crease.

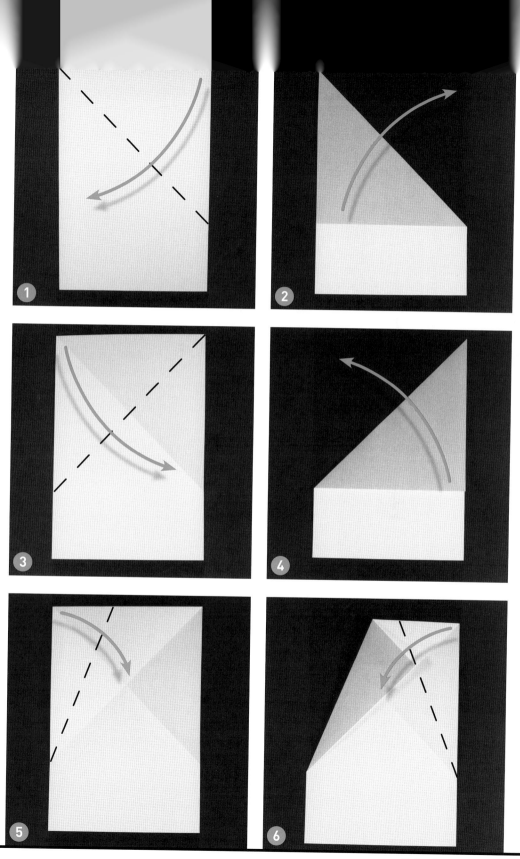

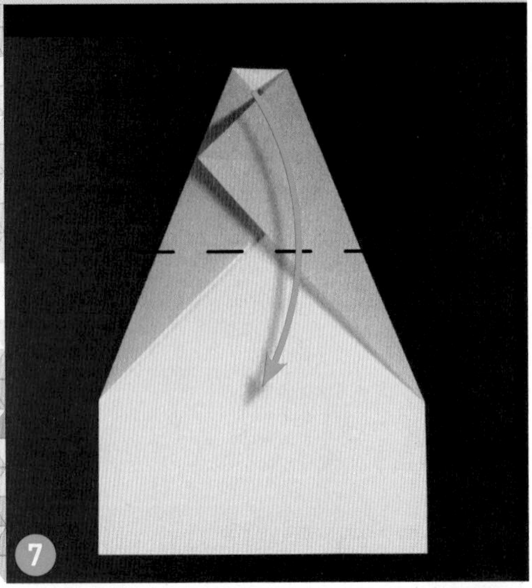

7

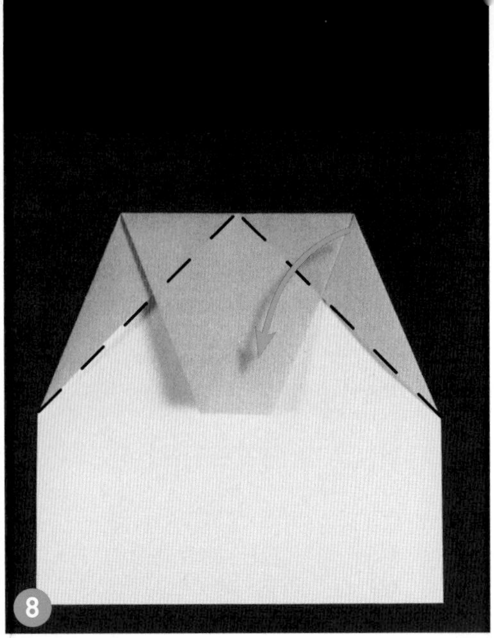

8

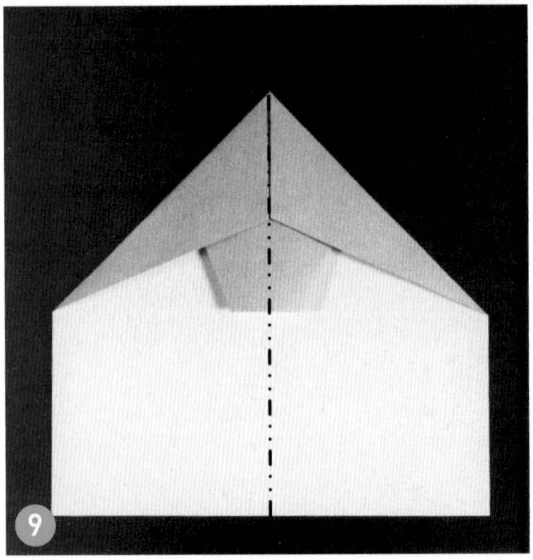

9

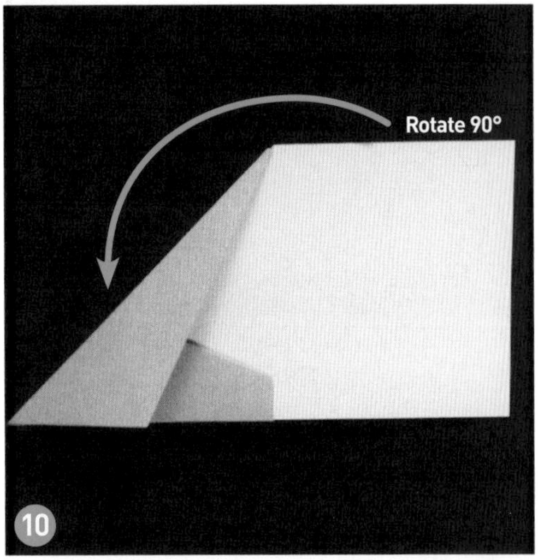

Rotate 90°

10

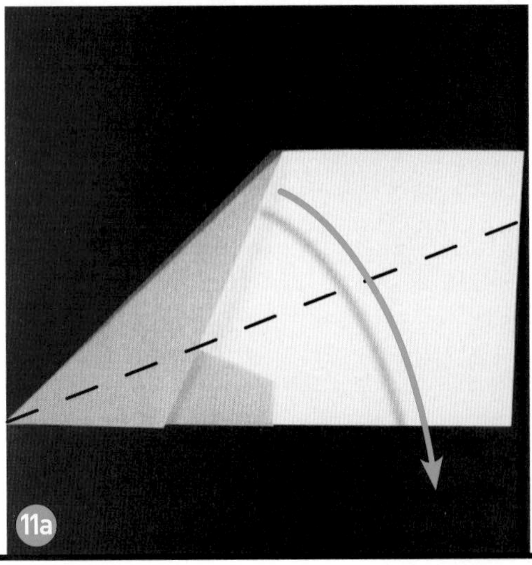

11a

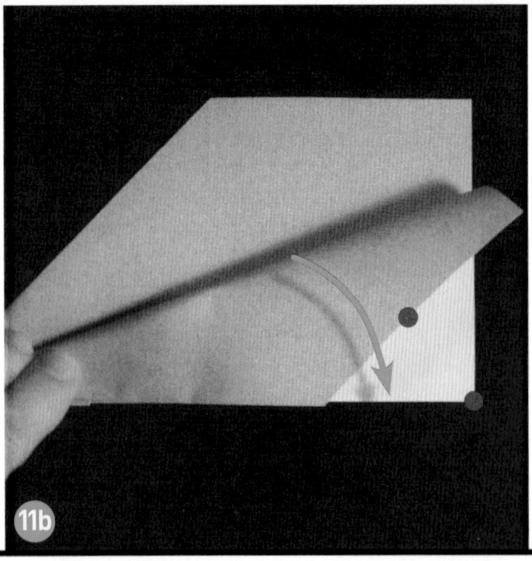

11b

7. Fold the top down, across the point where the diagonal creases cross.

8. Fold the top corners down to meet at the centerline. The top layer creases should line up with the bottom layer. Follow the existing creases.

9. Fold the plane in half. The mountain fold means you flip it over first, then fold in half.

10. Rotate the plane ¼ turn so that the center crease is on the bottom.

11. Make the wing fold. Start by positioning the creased edge of the wing against the center crease, but don't fold yet.
   Keep pulling the wing down until the little white triangle is gone. Now fold.
   Then make the other wing match.

12. Cut the tape into strips 30mm long: 3 strips about 2.25mm wide, and 7 narrow strips about 1.5mm wide.

13. Apply tape strips in the numbered order shown here. Note that strips 3, 4, 5, 15, and 16 are wider strips. Strips 3, 4, and 5 are all cut from one wide strip. Strips 10, 11, and 12 are all cut from one narrow strip.

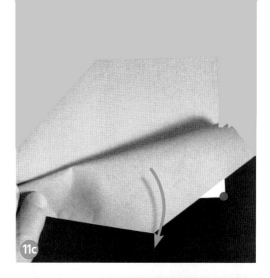

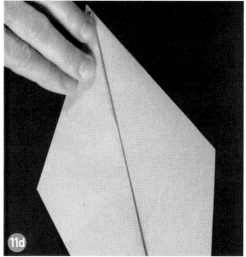

11c

11d

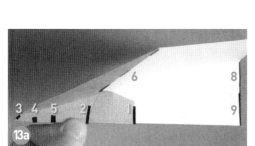

13a

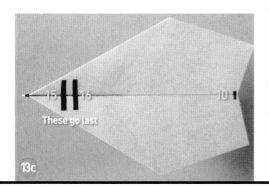

These go last

13c

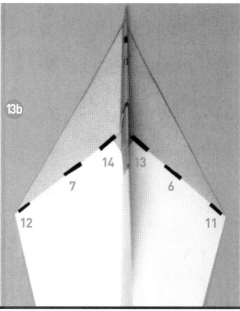

13b

14a

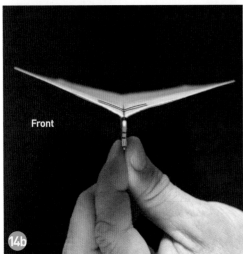

Front

14b

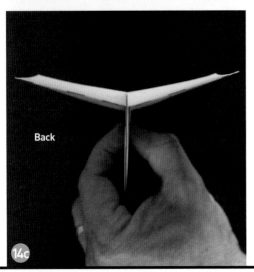

Back

14c

14. Use your dihedral gauges to set the angle at the nose at 165°, and at mid-wing 155°.

GOING THE DISTANCE

To throw your world record plane, hold it where the most layers meet (at the thickest part). Keep the wings level. Strive for smooth acceleration.

$1,000 Reward!

Use this design and officially break the world record for distance, and a $1,000 reward is yours. You must be named by Guinness as the world record holder.

Get the plane flying straight, and then work up to a fast throw. For turns at regular gliding speeds, you'll adjust the trailing edge of the plane normally See "The Flight Stuff" on the opposite page for tips on adjusting your plane's control surfaces.

But keep in mind that the faster this plane flies, the closer to the nose your adjustments will be made. At our highest throwing speeds, if the plane veered right on launch, I would bend the leading edge on the left down a little, at the nose.

World record throw by Joe Ayoob as the author (left) looks on.

John Collins teaches the Bluprint class "Plane Games: Make & Fly Paper Airplanes" where he shows how to fold and fly five of his best designs, including this world-record plane, and tests each one for distance, speed, and more. mybluprint.com/playlist/11671

The Flight Stuff

Know your aircraft parts. Notice that all the control surfaces are at the trailing edge.

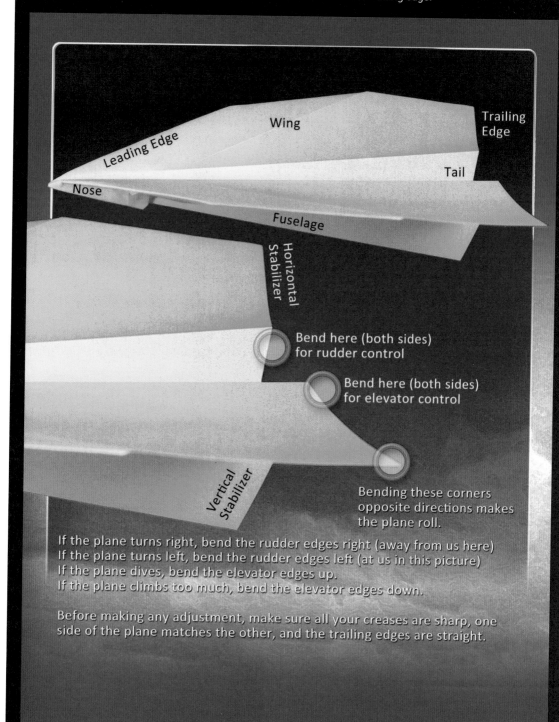

Wing

Trailing Edge

Leading Edge

Tail

Nose

Fuselage

Horizontal Stabilizer

Bend here (both sides) for rudder control

Bend here (both sides) for elevator control

Vertical Stabilizer

Bending these corners opposite directions makes the plane roll.

If the plane turns right, bend the rudder edges right (away from us here)
If the plane turns left, bend the rudder edges left (at us in this picture)
If the plane dives, bend the elevator edges up.
If the plane climbs too much, bend the elevator edges down.

Before making any adjustment, make sure all your creases are sharp, one side of the plane matches the other, and the trailing edges are straight.

Written and photographed
by William Gurstelle

Ismail al-Jazari
and the
Elephant
Clock

Build the amazing, accurate water clock invented by Baghdad's Golden Age engineer

When invading Germanic tribes deposed the last Roman emperor, Romulus Augustus, in 476 AD, Western Europe fell into a period of intellectual stagnation. The next five centuries were, at least in terms of the progress made in European-based science and technology, a very dark age. But the light of science was not snuffed out; instead, it found new places to shine, namely, China, India, and the Arab empires.

The 8th century saw the city of Baghdad quickly grow from a small settlement on the banks of Tigris to a military, commercial, and scientific powerhouse. In fact, Baghdad soon became one of the world's most important centers for education, culture, and the exploration of science.

The rulers of Baghdad were the caliphs of the Abbasid dynasty (750–1258), and some of them had a deep and abiding interest in technology. They invested a substantial amount in scientific explorations of many kinds.

The best remembered Arabic engineer of that time was Ismail al-Jazari, who was born sometime in the mid-12th century and died in 1206. He is known primarily through his landmark book, *The Book of Knowledge of Ingenious Mechanical Devices*. Published just before his death, its pages are filled with a plethora of descriptions of ingenious mechanical devices, along with *Make:* magazine-like instructions showing how to construct them. Included there are how-to advice and illustrations describing about 100 robotic automata, water fountains, water pumps, and clocks. Like Leonardo da Vinci, who lived 300 years after him, al-Jazari was a polymath adept at many things, and like da Vinci, he is widely remembered for sketches and notebooks of ahead-of-his-time ideas and gadgets.

TIME REQUIRED:
1–2 Weekends

DIFFICULTY:
Intermediate

COST:
$75–$99

MATERIALS

STRUCTURE:
» PVC pipe, white, 3" diameter, 20" long
» PVC pipe, clear: 4" diameter, 12" long, and 3" diameter, 8" long
» PVC couplings, white, smooth to smooth: 3" diameter (1) and 4" diameter (1)
» PVC reducer couplings, white, 4"-to-3" (2)

ADJUSTABLE FLOW SPOUT:
» Rubber stopper, #0 size
» Copper tubing, ¼" O.D., 3½" long

AUTOMATIC FLOW CONTROL VALVE:
» PVC pipe, white, ½" diameter, 4" long
» Balsa wood circle, 2¾" diameter, 1" thick You can easily cut this from a square piece using a coping saw.
» Foam rubber circle, 2¾" diameter foam rubber circle, ¼" thick I used scissors to cut up an old mouse pad.
» DWV pipe fittings, inset test cap, 3" diameter (2) These are available in the plumbing section of all big box and most hardware stores.

TOOLS
» Rubber mallet
» Electric drill and drill bits
» Hammer
» Hot glue gun and waterproof hot glue
» Epoxy filler such as J-B Weld WaterWeld Epoxy Putty Stick
» PVC primer and cement
» Coping saw or bandsaw
» Copper tube bender (optional) helpful but not 100% necessary

WILLIAM GURSTELLE's book series *Remaking History*, based on this magazine column, is available in the Maker Shed, makershed.com.

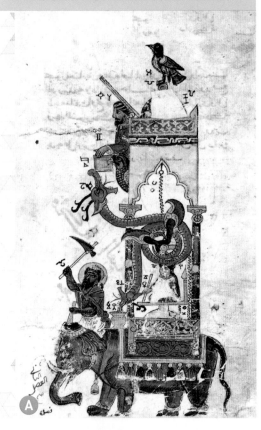

A

B

Modern biographers credit al-Jazari with inventing, refining, or at least anticipating many important technologies. Examples include sandcasting, valve seats, a windproof lamp, plywood, and an automated water flow regulator.

Historians place al-Jazari in the front rank of scientists and engineers from his era. His books are so detailed that many of his devices have been reconstructed by modern makers working from the drawings they contain. Some, like his famous Elephant Clock (Figure **A**), are so remarkable that several reproductions have been made and currently amaze visitors in museums and other public areas around the world (Figure **B**).

In this issue of Remaking History, we create an al-Jazari-inspired water clock. Like the original, our Elephant Clock has two interesting features that al-Jazari either invented or improved upon (Figure **C**). The first feature is an *automatic float valve* that keeps the water level in a tank constant, thus allowing a constant flow rate of water out of the tank. Normally, the rate of water leaving a vessel through a hole in the vessel's side or bottom slows down as the water level in the tank drops. This phenomenon, known as *Torricelli's Law*, is a big problem for water clocks because they indicate the passage of time by showing the difference in the height of water levels from one timing mark to another. Such nonlinear flows make measuring elapsed time very complicated. But with al-Jazari's automatic valve, the water flow in or out of the measuring tank is always constant.

The second interesting feature is the Elephant Clock's ability to adjust the rate of flow out of the tank. In al-Jazari's time, the length of an hour was determined by dividing the time between sunup and sundown into 12 equal parts. So, the length of an hour varied greatly by the time of year. Al-Jazari's Elephant Clock had an *adjustable flow spout* that could be easily adjusted to run faster or slower by rotating the curved spout to change the clock's *pressure head* (the height of the water in the tank above the orifice from which it flows out).

MAKE YOUR AL-JAZARI-INSPIRED WATER CLOCK

1. MAKE THE ADJUSTABLE FLOW SPOUT

1a. Drill a ¼" hole in the center of the #0 rubber stopper.

1b. Use the hammer to crimp one end of the copper tube, leaving only a very small orifice for the water to exit. Now, bend the copper tube into the shape shown in Figure **D** using a tube bender.

> **TIP:** If you don't have a tube bender, you can fill the copper tube with salt, plug the ends, and form the tube by carefully bending it with your hands.

1c. Insert the bent tube into the rubber stopper hole (Figure **E**).

2. MAKE THE AUTOMATIC FLOW CONTROL VALVE

2a. Hot-glue the foam rubber circle to the balsa wood disc.

> **TIP:** You can use a few ½"-long nails to attach the foam rubber to the wood if you have trouble gluing it.

2b. Fill one side of a DWV test cap with mixed epoxy putty and allow it to dry (Figure **F**). Drill a ⅞"-diameter hole in the center of the cap. Insert the ½" diameter (ID) pipe into the hole, extending one end 2¼" past the bottom. Square the pipe to the bottom of the test cap and hot-glue the pipe into place.

2c. Use the PVC primer and cement to attach the test cap/pipe assembly to the inside of one of the 4"-to-3" PVC reducing fittings. Make sure the 2¼"-long end of the pipe is on the side with the 3" opening (Figure **G**). Add a bead of hot glue to the connection to ensure it is watertight.

2d. Drill a ⁹⁄₁₆" hole for the stopper on the side of the 3"-diameter clear PVC pipe, 2¼" from the bottom. Drill a ¼" vent hole in the same pipe, 2½" from the top.

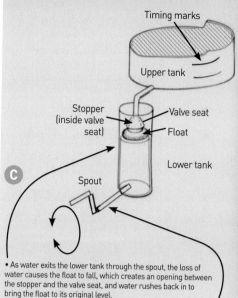

Timing marks
Upper tank
Stopper (inside valve seat)
Valve seat
Float
Lower tank
Spout

C

• As water exits the lower tank through the spout, the loss of water causes the float to fall, which creates an opening between the stopper and the valve seat, and water rushes back in to bring the float to its original level.

• The curved or L-shaped spout can be rotated up or down. When it's rotated up, the pressure head is smaller and flow is smaller. When rotated down, the head is larger and flower is greater.

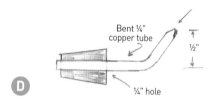

Bent ¼" copper tube
½"
D
¼" hole

E

F

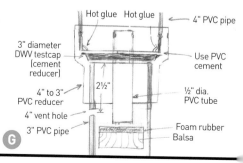

Hot glue   Hot glue   4" PVC pipe
3" diameter DWV testcap (cement reducer)
Use PVC cement
2½"
4" to 3" PVC reducer
½" dia. PVC tube
4" vent hole
3" PVC pipe
Foam rubber
Balsa
G

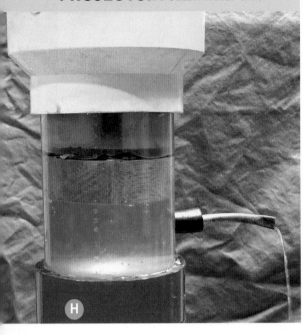

2e. Use the PVC primer and cement to attach the remaining test cap to the inside of the 3" white PVC smooth coupling. Add a bead of hot glue to the connection so it's watertight. Figure **H** shows how the float valve will look in action.

3. ASSEMBLE THE WATER CLOCK

3a. Assemble the water clock by connecting the PVC fittings, wood float, and adjustable spout as shown in the assembly diagram (Figure **I**). Use the rubber mallet to seat the fittings onto the pipe. Note that only two joints are cemented together using the PVC primer and cement. The others must be able to be opened for troubleshooting and cleaning.

3b. Optional: You can add some color to your project by using dye to stain the white PVC pipe and fittings. For tips, see Sean Michael Ragan's tutorial at makezine.com/projects/stain-pvc-any-color-you-like.

USING YOUR AL-JAZARI WATER CLOCK

To set up the water clock, *slowly* fill the 4"-diameter water reservoir pipe with water. As the top tank fills, water will flow into the second tank. As it does, the float will rise with the water. When the foam rubber contacts the bottom of the ½" pipe, the rubber seals off the pipe, preventing additional water from entering the second tank. But, as water exits the second tank via the spout, the level falls, unsealing the tube, and water flows again. This process repeats indefinitely as long as there is water in the top tank.

To gauge the passage of time, mark the water at the beginning by making a mark on the 4" clear PVC with a dry erase marker. Then use the marker to mark whatever time intervals you're interested in on the side of the top tank.

By rotating the spout up or down, you can control the rate at which the water exits and therefore the amount of time between marks. When the spout is up, water flow is decreased. When the spout is down, the pressure head is larger and the water flow increases. ◢

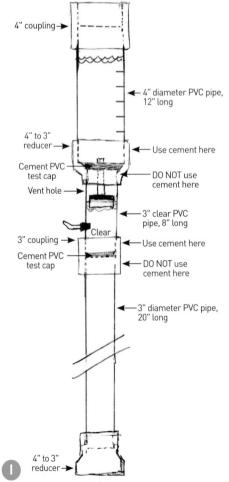

4" coupling →

← 4" diameter PVC pipe, 12" long

4" to 3" reducer →

← Use cement here

Cement PVC test cap

← DO NOT use cement here

Vent hole →

← 3" clear PVC pipe, 8" long

Clear

3" coupling →

← Use cement here

Cement PVC test cap

← DO NOT use cement here

← 3" diameter PVC pipe, 20" long

4" to 3" reducer →

HIROSHI MAEDA is a mechanical designer from Shiga Prefecture, Japan.

Tiny Dancer

With a Raspberry Pi Zero as its brain and gutted servos for muscles, this homebrew robot doll is a stunningly smooth mover

Written and photographed by Hiroshi Maeda

When I was young, I liked mechanical and electronic work, and when I was in elementary school I was crazy about BASIC programming. Since then, making things has been part of my everyday life. Currently, I am working on mechanical designs for my ideal machine form.

About 15 years ago I designed and manufactured my first robot, called "Jim Spartan" (Figures 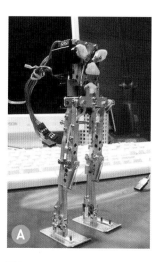 and 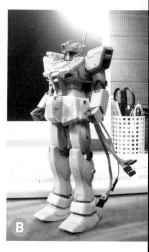). It was a very small humanoid robot with a height of 180mm (7.09") and a weight of 195g (0.43lb). By combining multiple link mechanisms, it could walk slowly and quietly, using four R/C servos to move the 10-axis legs and two-axis arms. It had an H8 microcomputer for the controller. I am very nostalgic about it.

ST-01 is my latest finished robot (Figure ). It is a 495mm (19.49"), 1150g (2.54lb) doll-shaped robot that moves smoothly, can salute, dance, and can even walk on two legs. The previous version, ST-00, even completed the running competition at Robo-One (watch here: makezine.com/go/st-00).

I started the conceptual designs for ST-01 in August 2015 and completed the first model in December 2017. I've shown it at the Ogaki Mini Maker Faire 2018, Maker Faire Kyoto 2019, and Maker Faire Tokyo 2019, where people complimented its smooth, human-like movement.

A

B

C

1

2

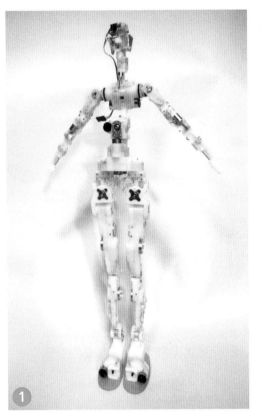

3

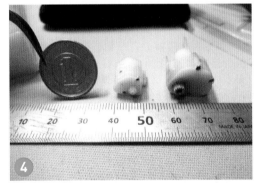

4

1. CNC-milled frame with servomotor internals installed.
2. Unfinished vacuum-formed body shell mounted.
3. A Raspberry Pi Zero and RX220 microcontrollers control the servos and movement.
4. Size comparison of the rebuilt servos.
5. Testing servos with the microcontrollers.
6. Painting the vacuum-formed shell.
7. Wiring the mounted microcontrollers to the motors.
8. Partially assembled ST-01, with eyes and hair in place.

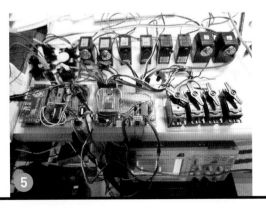

5

The internal mechanical design of the ST-01 is a serial link mechanism that is very common in humanoid robots. However, using a regular R/C servo for a two-legged walking robot does not lend itself to creating a beautifully proportioned robot. To overcome that, I take the gear, motor, and printed circuit board out of a standard servo and reassemble the mechanical parts into the robot's body, which I also designed and CNC-routed. I then use an RX220 microcontroller to control the rebuilt servos, and a Raspberry Pi Zero to generate their movement. For the body shell, I vacuum-form each piece from my hand-carved wood prototype. In other words, I design and manufacture every part, from the top of the head to the tip of the foot.

Mechanical design is my profession, but I also like control design just as much, so I do both by myself. I think that in order to have a good mechanical design, it is necessary to be able to do basic control design — and the converse applies as well. For programming I like C and C++.

ST-01's motion is generated with the Raspberry Pi Zero. The time required depends on how the robot is moved. At the moment, software development is in progress, and it takes a certain amount of time because manual operations such as animation frame division are required.

The production of ST-01 is my hobby, so the design and production processes have many compromises. There are plenty of ideas that I really want to implement, but the realization will require the cooperation of companies. However, I want to improve things in the next iteration, as I can still do more as an individual person. A smaller, high-performance female robot ST-02 is currently under production (I've completed the machine design). In particular, the body has greatly improved so it will be created like a commercially available doll body.

You may notice that the female robots that I design and produce only have a model number and do not have human-like names. There is a reason for this: In the future, when commercialization is realized, the owner can give it its name. I do not know when it will happen, but I will never give up. Please wait patiently if you want to become the owner of a "female robot ST."

You can follow the progress of my robots at twitter.com/med_mahiro. ◐

Check out ST-01's flowing dance moves: makezine.com/go/st-01

Storm Warning

Build your own wearable lightning detection system!

Written and photographed by Alex Wulff

This personal **Lightning Detector** is exactly what it sounds like: a small device that alerts you to nearby **lightning strikes.** It even tells you how far away they are. The materials for building it cost less than a commercial lightning detector, and you'll get to hone your circuit-making skills in the process.

This project is based upon the AS3935 lightning sensor IC, with a carrier circuit from DFRobot. It detects electromagnetic radiation that's characteristic of lightning and uses a special algorithm to convert this information to a distance measurement.

The sensor can detect lightning strikes up to 40km (25 miles) away and is capable of determining the distance of a strike to within a tolerance of 4km (2.5 miles). While this is a reliable sensor, you should not depend upon it to warn you of lightning strikes if you're outdoors. Your own circuit handiwork may not be as reliable as a commercial lightning detector.

1. Plan out the circuit

As there's a relatively small number of parts in this build, the circuit is not particularly intricate. The only data lines are the SCL and SDA lines for the lightning sensor and one connection for the buzzer. A lithium ion polymer battery powers the device, so I also decided to integrate a LiPo charger into the circuit.

Figure Ⓐ depicts the entire circuit. Note that the connection between the LiPo battery and the LiPo battery charger is via the JST male/female connectors and does not require soldering.

TIME REQUIRED:
3–4 Hours

DIFFICULTY:
Intermediate

COST:
$40–$60

MATERIALS
» **Beetle microcontroller** DFRobot #DFR0282, dfrobot.com/product-1075.html. It's a very small Arduino Leonardo board.
» **Gravity: Lightning Distance Sensor** DFRobot #SEN0290, dfrobot.com/product-1828.html
» **Lithium battery charger** DFRobot #DFR0208, dfrobot.com/product-851.html
» **LiPo battery, 500mAh** Amazon #B00P2XICJG
» **Piezo buzzer, 5V** such as Amazon #B07GJSP68S
» **Slide switch, small**
» **Hookup wire** solid or stranded

TOOLS
» **Computer with Arduino IDE** free at arduino.cc/downloads
» **Soldering iron and solder**
» **Hot glue gun**
» **Wire strippers**
» **3D printer (optional)**

Adobe Stock - TAW4

ALEX WULFF is a maker and student in electrical engineering at Harvard. He enjoys creating projects and helping others learn about electronics, technology, radio, space, and more. For more projects, and his new book teaching makers about radio communications, visit alexwulff.com.

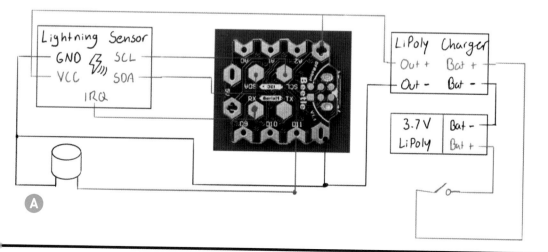

Ⓐ

- Positive (+) on the lightning sensor to positive (+) on the Beetle
- Negative (–) on the lightning sensor to negative (–) on the Beetle
- Clock pin (C) on the lightning sensor to the SCL pad on the Beetle
- Data pin (D) on the lightning sensor to the SDA pad on the Beetle.

Also connect the IRQ pin on the lightning sensor to the RX pad on the Beetle. The connection needs to go to a hardware interrupt on the Beetle, and the RX pad (pin 0) is the only interrupt-capable pin remaining.

WIRE THE BUZZER
Connect the buzzer's short lead to the negative (–) terminal on the Beetle (ground), and its long lead to pin 11. Since the buzzer's signal pin should be connected to a PWM pin for maximum versatility, pin 11 is perfect.

ADD THE SWITCH
Add a switch inline to the battery to turn the project on and off. First, solder two wires to adjacent terminals on the switch. After soldering, I fixed these in place with hot glue, as the switch's connections are fragile.

Cut the red wire on the battery about halfway down and solder the wires from the switch to each end. You can see these connections on the righthand side in Figure B. Make sure you cover the exposed sections of wire with heat-shrink tubing or hot glue, as these could easily come in contact with one of the ground wires and make a short. After adding the switch, you can plug the battery into the battery charger.

FOLD EVERYTHING IN
The last step is to take the gangly mess of wires and components and make it look somewhat presentable (Figure C). This is a delicate task, as you want to be sure you don't break any wires. Start by hot-gluing the LiPo charger to the top of the LiPo battery, then glue the Beetle on top of that. Finally, glue the lightning sensor at the very top. I left the buzzer to sit off to the side, as shown in Figure C. The final result is a stack of boards with wires running throughout. I also left the switch's leads free to integrate into a 3D-printed case later.

2. Assemble the circuit
This device is a great candidate for the circuit assembly technique known as *free-forming*. Rather than affix the parts to a substrate such as a perf board, you can just connect everything with wires instead (Figure B). This makes the project much smaller and faster to assemble, but can produce less aesthetically pleasing results. I like to cover my freeform circuits with a 3D-printed case at the end. You can see how I assembled my freeform circuit for this project in the video at makezine.com/go/lightning-detector.

CONNECT BEETLE TO CHARGER
Unsolder the green terminal blocks from the LiPo charger. You don't need them, and they take up space. Then connect the positive (+) and negative (–) terminals of the LiPo charger to the positive (+) and negative (–) terminals at the front of the Beetle. This feeds the raw voltage of the LiPo battery straight into the microcontroller. The Beetle technically needs 5V, but it will still operate on the roughly 4V from the LiPo.

WIRE THE LIGHTNING SENSOR
Cut the included 4-pin cable so that roughly 2" of wire remains. Strip the ends and plug the cable into the lightning sensor, making the following connections:

3. Program the microcontroller

Open the Arduino IDE on your computer and make sure the Tools→Board selection is set to "Leonardo." Download and install the library for the lightning sensor at github.com/DFRobot/DFRobot_AS3935. Then download the project code from alexwulff.com/files/lightningcode.zip and upload it to the Beetle. The software is simple but it's very customizable to suit your needs.

When the device detects lightning, it will first beep many times to alert you that lightning is nearby, then beep a certain number of times corresponding to the lightning's distance in kilometers. If lightning is less than 10km (6.2 miles) away, the device will emit one long beep. If it's more than 10km (6.2 miles), the device will divide the distance by 10, round it, and beep that many times. For example, if lightning strikes 26km (16 miles) away, the device will beep three times.

The software revolves around interrupts from the lightning sensor. When the lightning sensor detects the electromagnetic radiation from a lightning strike, it sends the IRQ pin high, which triggers an interrupt in the microcontroller. The sensor can also send interrupts for non-lightning events, such as if the interference/noise level is too high. If so, you'll need to move your detector away from any electronic devices. Radiation coming from these devices can easily dwarf the comparatively weak radiation from a distant lightning strike.

4. 3D print a case (optional)

I modeled a case for my device. You can download the files for 3D printing from thingiverse.com/thing:3769758 (Figures D and E). The top of the case snaps onto the bottom, so no special hardware is required. I made it big enough that it will probably fit your device (Figure F), but you can also try creating your own case:

- Measure the basic dimensions of your device
- Model your device in a CAD program (I like Fusion 360 — students can get it for free)
- Create a case by offsetting a profile from the device model. A tolerance of 2mm generally works well.

Detecting Lightning Strikes

Congratulations, you now have a fully functioning lightning detector! I recommend waiting until there's a thunderstorm around you to make sure the device actually is capable of detecting lightning. I don't know the reliability of this sensor, but mine worked on the first try.

Charging the device is simple — you can just plug a micro-USB cable into the LiPo charger until the charging light turns green. Make sure the device is switched on while you charge it, or no power will go to the battery!

Making Modifications

There are many modifications you can make to the software to make your lightning detector more useful and user-friendly.

- Change the beeps: You can use the *Tone.h* library to generate more pleasant-sounding notes.
- Implement sleep mode: The ATmega32u4 microcontroller (the chip powering the Beetle) supports hardware interrupts in sleep mode. You can place the device into sleep mode, then have it wake up when it receives an event from the lightning sensor. This should greatly extend the battery life of the system. ◎

Watch the full video tutorial at makezine.com/go/lightning-detector

Written and photographed by Bob Knetzger

Custom Toast Stamper
Emboss your breakfast like a boss!

TIME REQUIRED:
1–2 Hour

DIFFICULTY:
Easy

COST:
$10–$20

MATERIALS
» **Acrylic sheet, ¼" thick, about 6"×12"** Extruded acrylic is easy to laser-cut and to cement. Cast acrylic might be a bit less sticky to cut by hand.
» **Acrylic cement, methyl chloride solvent type**

TOOLS
» **Laser cutter, bandsaw, or hand coping saw**
» **Small brush or capillary needle applicator** for the solvent cement
» **3D printer (optional)** You can also print this project instead of cutting it.

BOB KNETZGER is a designer/inventor/musician whose award-winning toys have been featured on *The Tonight Show*, *Nightline*, and *Good Morning America*. He is the author of *Make: Fun!*, available at makershed.com and fine bookstores.

Here's a "toast" to makers everywhere! This easy project makes a fun, custom stamper that embosses your design onto a piece of toast. It's a simple 3D form that can be made by hand with minimum equipment, or if you have access to a laser cutter, you can knock it out in just a few minutes. It also lends itself to 3D printing.

The version shown makes cute Makey robot-imprinted toast but you can create your own custom design with any message or image.

CUT YOUR OWN TOAST STAMPER
To make your stamper by hand, glue the pattern onto a piece of ¼" acrylic. Use the full-size pattern here (Figure Ⓐ) or go online to makezine.com/go/toast-stamper for a PDF.

Cut out the bread slice backing piece and the other smaller parts for the Makey bot shapes. You could use a coping saw or bandsaw. Take your time, as the acrylic can be brittle. If you break a corner, no worries: You'll be solvent bonding the parts together onto the backing later anyway. Any gaps or cracks won't show up in your final stamped bread.

I cut my parts on a Glowforge laser cutter from Proofgrade Thick Acrylic (Figures B and C). Or use your own material with Full Power and 1000 Speed settings. Online you'll also find a *.ai* vector art file as well as a *.svg* cutting file.

Solvent-bond the parts together with methyl chloride acrylic cement in a well-ventilated area (Figure D). Be sure to let the bonds set fully before using the stamper with food.

OR 3D PRINT IT

The design is so simple that it could be 3D printed as well. In Tinkercad or your favorite 3D design program, extrude the 2D shapes in the *.svg* file, then join them together. Orient the shapes so that the backing plate is on the bottom and printed first.

GET CREATIVE

Make a personalized version for a special birthday, Valentine's, or any personalized message. Emoji toast, anyone?

To make your own design, use the bread slice backing form as the basic size. Also remember to "mirror flip" your design so that any text reads "right" on the final stamped bread.

NOW YOUR TOAST

To use your stamper, just press your design down firmly onto a slice of bread, then toast!

Soft white bread works best: It compresses down well with good "resolution" (Figure E). The compressed, denser areas stay white while the other areas toast nicely. ⊘

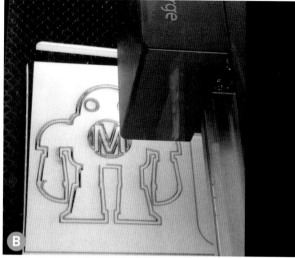

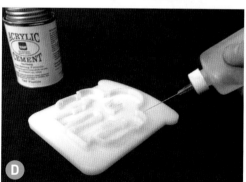

Fasten-ating World of
FASTENERS

Get a grip on the nuts and bolts of holding your project together with this guide

Written by Brian Bunnell and Samer Najia

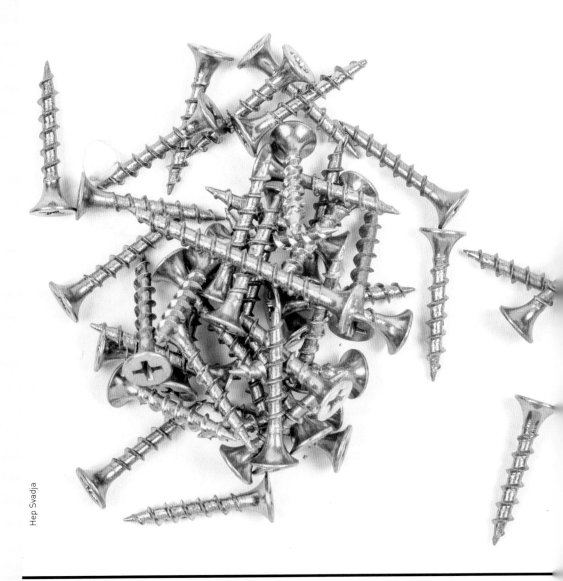

Hep Svadja

In the manufacturing phase of a project build, you will need to provide a means by which to fasten parts and components together. This is where fasteners and adhesives come into play. For the purpose of this discussion, we will classify *fasteners* as physical devices used to attach or fasten materials, parts, and components together in a non-permanent way. *Adhesives* will be generally defined as a liquid or thin film that bonds materials or parts in a more permanent fashion.

Types of Fasteners

Fasteners come in all sorts of sizes, types, and materials and offer functionality that goes well beyond the proverbial nuts-and-bolts way of keeping things together. Consider the simple screw: nothing more than a spiral wrapped around a shaft that is either tapered or flat at one end, and a head at the other that can tightly hold two or more parts together. Fasteners are available in both metric and Imperial (English) units of measure, which can sometimes cause confusion. We could write several chapters on all the various fasteners you see just around you on a daily basis, but in this chapter we'll cover the fasteners that we, as makers, have found to be most useful for a wide range of projects.

We're sure you will find the array of fastener types quite dizzying. Fasteners have been around for millennia and run the gamut from the ridiculously simple to seriously complex.

BRIAN BUNNELL is a mechanical engineer by education but a maker at heart. He earned his engineering degree from Clemson University in 2000 and has been working in mechanical design ever since. Brian began making early on (creating crazy projects with his Dad), and making quickly became his lifelong passion.

SAMER NAJIA holds a degree in mechanical engineering from Duke University but he is a serial maker and building things is his true passion. Samer spends countless hours building progressively larger and more complex projects, disappearing into his garage or loft for hours. Some of these projects are outlandish, but that just means they need more design work.

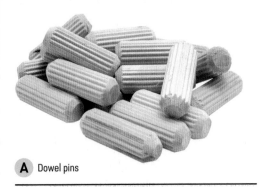

A Dowel pins

B Nails

C Pop rivet

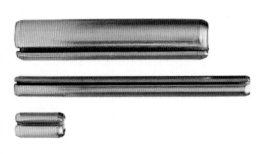

D Pins

- **Dowels** (Figure **A**): These are pegs or pins without a head that makers can pound into a hole. Wooden dowels, for example, when coated in wood glue, expand into the hole as the glue cures, and secure parts together very tightly. Dowels are typically made of wood, and sometimes have longitudinal ridges ("flutes") that dig into and "grip" the sides of the hole, trapping glue in between them.

- **Nails** (Figure **B**): Nails have a tapered end that forces apart the material into which they are being driven. The material then contracts around the nail, holding it in place. The material exerts a greater force on the nail as the nail is driven deeper.

- **Rivets** (Figure **C**): You will mostly encounter "blind" rivets. Also known as "pop" or "pull" rivets, they have a shank that breaks as you pull them with a special riveting tool, thus securing the parts by compressing and deforming the end of the shank. These come in pan head (a round, domed head) or countersunk (v-shaped head) types. You might also encounter plastic rivets that can be put in and taken out by hand but these are really "low" grip fasteners.

- **Pins** (Figure **D**): These are similar to dowels, however, they are typically made of metal. A common example of a pin is a **roll pin**. A roll pin is typically made from flat spring steel rolled into a cylinder. When a roll pin is driven into a hole, the pin collapses slightly, thereby jamming itself into the hole.

SCREWS

Screws can generally be divided into three types; wood screws, sheet metal screws and machine screws. Screws with tapered threads are designed to screw into a softer material, while screws with straight threads are designed to use with a nut, or thread into a matching threaded hole. Additionally, some wood or sheet metal screws are designed with a small drill-bit type point at the end, and referred to as "self-tapping" or "self-drilling" screws.

- **Wood screws** (Figure **E**) have tapered threads, and are designed to be threaded into a plain hole in wood, so that the wood deforms around the threads of the screw to hold it in place. They generally have a non-threaded portion near the screw head that is intended to fit loosely in a slightly oversized hole in one layer of material, with the threaded part in a second layer, holding the two layers together. Wood screws also frequently have a flat head intended to fit into a countersunk hole in wood (or designed to make such a hole as the screw is driven into the wood). Larger sizes of wood screws, used for heavy structural connections, are called **lag screws** or even **lag bolts**.

- **Sheet metal screws** (Figure **F**) also have tapered threads, and are designed to be threaded into softer material, allowing the material to deform and hold the screw tightly in place. Sheet metal screws look similar to wood screws, but are made to hold much thinner layers of material together. They tend to have smaller threads, and they usually have a round or hexagonal head that stands proud of the material they're driven into. Also like wood screws, sheet metal screws are not used with nuts or pre-threaded holes.

- **Machine screws** are designed to be threaded into a nut or a threaded hole in another part. Their threads are not tapered. Larger sizes of machine screws are called **bolts** (Figure **G**).

- **Carriage bolts** (Figure **H**) have a domed head with a square portion below the head. This type of bolt is designed to be used in applications where a smooth, low head is required for aesthetic and/or safety reasons. An example of when this type of bolt is used is when building children's playground equipment. If a child brushes up against the smooth head of a carriage bolt he/she will not hurt themselves. When a carriage bolt is used to secure wooden parts, the square

E Deck screw

F Self-drilling sheet metal screw

G Hex head bolt

H Carriage bolt

I Shoulder bolt (also called a stripper bolt)

J Eye bolt (left), eye lag (right)

K Hanger bolt

L Hex nut

M Nyloc nut

N T-slot nut

portion of the bolt is designed to be pulled or driven into a clearance hole drilled for the bolt. The square feature "bites" into the material around the clearance hole thereby keeping the bolt from rotating while the nut is tightened on the other end. This type of bolt can also be used to secure metal parts together. For this type of application, the top layer of metal to be secured generally will have a square hole to receive and engage the square part of the bolt.

- **Shoulder bolts,** also known as **stripper bolts** (Figure **I**), have a very precise shank or smooth portion of the bolt intended to be used as a pivot. The end of the bolt is threaded with a thread diameter less than that of the shank. This smaller size thread creates a step at the bottom of the bolt that the bolt seats against when it is tightened down to the face of a material.

- **Eye bolts** and **eye lags** (Figure **J**) have a circular ring on one end and are threaded on the other. Eye bolts have a machine screw thread, while eye lags have a wood screw thread. This type of fastener is used for securing rope or chain.

- **Hanger bolts** (Figure **K**) have a machine screw thread on one end and a lag or wood screw thread on the other. This type of fastener is used when a machine screw threaded stud is required to protrude from a wooden part.

NUTS
Like screws, there are plenty of variations:

- **Regular nuts** are typically made of metal with a threaded hole running through the middle. The outside of the nut has a specific geometry by which the nut is intended to be gripped and rotated. It is common for the outside geometry to be a hexagon, i.e. a **hex nut** (Figure **L**), but you will occasionally come across nuts that have a square outside geometry.

- **Lock nuts,** sometimes called **nyloc nuts** (Figure **M**), are nuts that have a piece of nylon (typically) embedded in the hole that grabs the threads of a screw and prevents the nut from working itself loose over time. This type of nut is quite handy for applications where a bolt should be mechanically secured to a component, but not completely locked in place. A typical example of this is when using a bolt as a pivot for something like a lever. You don't want to completely lock the lever down with the bolt used as a pivot or axle, but you also don't want the lever to become loose or unsupported. This is where a nyloc nut is quite useful. Using a nyloc nut, the bolt being used as a pivot can be secured in place without being fully tightened, thereby allowing the lever to rotate about the bolt.

- **T-slot nuts** (Figure **N**) are similar to regular nuts in that they are also typically made of metal and have a tapped hole through the middle. However, the outside geometry of a T-slot nut differs significantly from that of a regular nut. The stepped outside shape of this type of nut is designed to engage in a part or material specifically designed to accept it. Aluminum extrusion is one example of a material designed to accept T-slot nuts, as shown in Figure **O**.

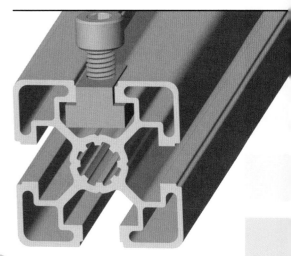

O Aluminum T-slot extrusion with T-slot nut

P Castle nut

Q Rivnut

R Installed rivnut

S T-nut wood insert

T Thermoplastic insert

U Plain washer

V Lock washer

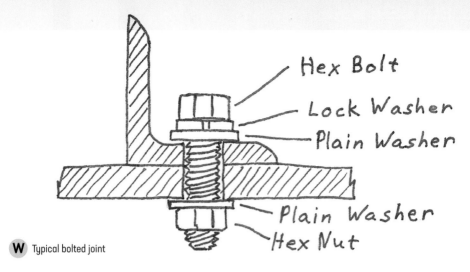

W Typical bolted joint

- **Castle nuts** (Figure **P**) have notches in them so that a pin (commonly called a **cotter pin**) can pass through the notches and a hole in the screw shank. This prevents the nut from rotating and working itself loose.

- **Rivnuts** (rivet nuts) (Figure **Q**) are essentially rivets that have a threaded hole in them to accept a screw. Rivnuts are great for securing panels in sheet metal (Figure **R**).

- **Inserts** (Figures **S** and **T**) come in a variety of forms suited to specific materials. They're similar to rivnuts in that they provide a threaded hole, but can be put into plastics, wood, and composites. Inserts for wood are commonly called **T-nuts** (as opposed to T-slot nuts) and get hammered into a hole in the receiving material. **Thermoplastic inserts** go into acrylic and other plastics by heating the brass insert, which melts the material around it. Once cooled, this forms a solid mechanical bond between the insert and the plastic.

WASHERS

These are an important part of mechanical fastening systems. They are essentially a thin disc with a hole in the middle.

- **Plain washers** (Figure **U**) are typically used to spread the load that a bolt exerts on the material it is fastening together, in order to prevent the bolt head from deforming and/or pulling through the material.

- **Lock washers** are designed to help secure the bolted joint from working loose due to vibration. There are many types, but the most typical is a "helical" or "spring" lock washer (Figure **V**), which is essentially a plain washer that has been cut and sprung out of plane.

  When a bolt exerts a load on a lock washer it is forced to flatten out, causing the washer to dig into the material you are fastening. It also provides some axial force on the bolt by virtue of it being forced flat. In other words, the lock washer tries to spring back. The forces generated by both digging-in and spring-back help to keep the bolt from rotating loose.

  In most applications where a lock washer is needed, you will also want to use plain washers. Figure **W** shows a typical bolted joint that has both a lock washer and plain washers. Note that only one lock washer is required for a joint such as this. However, two plain washers are used to distribute the load over a larger area of the materials being joined. ◗

You can find this article and much more useful information in our new book *Make: Mechanical Engineering for Makers*, available soon at all major booksellers and on makezine.com/go/mechanical-engineering-for-makers/.

Quick and Easy
BOX JOINTS

Master this stylish joinery technique with a simple-to-make table saw jig

Written by David Picciuto

DAVID PICCIUTO makes weekly videos meant to inspire and teach woodworking. He is a firm believer that creativity is not something you're born with; rather, it is a muscle that can be strengthened. Based in Toledo, Ohio, he's written two books and has made hundreds of videos. Find more at MakeSomething.com.

Dan Struffolino

TIME REQUIRED:
1 Hour

DIFFICULTY:
Moderate. Safe table saw experience is recommended.

COST:
$50–$200 depending on blade selection

MATERIALS
» **Plywood, Baltic birch,** ¾" ×3½" ×14"
» **Solid hardwood** enough to make a box of your choosing
» **Wood glue or CA glue**

TOOLS
» **Table saw**
» **Flat-kerf joinery blade or dado stack**
» **Table saw sled or miter gauge**
» **Clamps, hand screw or F-style**
» **Sandpaper**
» **Sander, orbital or disc**

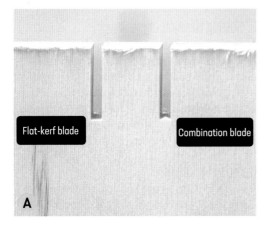

Flat-kerf blade   Combination blade

A

B

Box joints (sometimes called *finger joints*) are one of my favorite methods of joinery. They look absolutely beautiful, are extremely strong, and allow you to cut your boards to length without calculating for the size of the joint. For example if you wanted your final dimensions of your box to be 12" long and 9" wide you would cross cut your boards to 12" and 9". Box joints can be made with a router or table saw; in this example we'll make a dead-simple table saw jig and build a simple wooden box. You ready? Here we go!

> **NOTE:** You can also purchase jigs to help you make box joints but they are costly and involve setup time. I own one of those jigs and although it's fantastic at cutting variable-width joints it requires me to watch a video to relearn how to use it every time I break it out.

1. Getting ready
The first thing you need to do is determine the width of the fingers. A standard table saw blade is ⅛" wide, which will give you ⅛" fingers. If you're looking for larger fingers you will need a dado stack for your table saw. In this example we'll be using a single blade to make ⅛" fingers. It's highly recommended to use what's called a *flat-kerf joinery* blade (sometimes called a flat-bottom blade) with this approach. A rip, crosscut or combination blade will not leave a flat bottom (Figure **A**). Although they will work, your joints won't have that clean, classic look. I also like to make the height of my boxes in multiples of my finger width size. For example if I'm making ⅛" fingers I want the height of the box to be divisible by ⅛" so I'm not left with partial fingers.

Go ahead and cut 4 pieces of hardwood for your box sides (in the photos I'm using ¾" mahogany). Your box can be any size you want. Just remember the taller it is, the more fingers you'll have to make. 12" long, 9" wide and 2" high is a good first target. You will also want to cut 2 extra pieces to use for test cuts when setting up the spacing. These pieces can be any size, but 2" wide and 4" long out of hardwood should do the trick.

2. Kerf cutting
Next thing you need to do is cut a kerf into a piece of ¾" thick plywood for the jig (Figure **B**).

Dan Struffolino

The length and width of this piece aren't critical — mine here is 3½"×14". Set your blade to ½" above the table and run your plywood through the blade near the middle using your table saw sled or miter gauge with a backing board.

3. Cutting the pin

Now you'll need to cut a piece of hardwood to fit into that kerf you just cut (Figure C). This piece will be the *pin*. You'll cut it long at first: 8" long, ½" high and ⅛" wide. Getting that width can be kind of tricky and may take a couple of tries. It should fit into the kerf you cut in Step 2 without falling out. If you're having trouble getting that perfect width you can always cut oversize and sand it down to fit.

4. Cutting the pin to size

Take the pin and cut it off at 1½" (Figure D). The longer of the two pieces will be used as a spacer in Step 5.

Take the 1½" pin and glue it into the plywood kerf (Figure E), flush with the back. Make sure it's not sticking out the back of the plywood. You can use either wood or CA glue to secure it in place.

5. Setting the spacer

Next, take your board with the pin and place it up against your table saw sled or miter gauge with a backing board. Use the 6½" cutoff from the pin as a spacer between the blade and the pin (Figure F). Because this spacer is the thickness of your blade, it'll set your pin in near perfect placement for cutting the correct size fingers.

6. Clamping in place

Once you get the correct spacing, clamp your pin board to your sled or miter gauge (Figure G).

7. Sanding the pin

Use a piece of sandpaper to round over the top of the pin (Figure H). This will make your workpiece board slide in and out a lot easier. You may even want to rub a bit of wax on it to reduce friction.

8. Start finger cutting

Now it's time to make your first cuts! Move one of your test pieces up against the fence and against the pin. Run your board through the blade, continuing to hold it tightly (Figure I). If you don't feel comfortable holding the piece you can always clamp and unclamp the board after each cut.

G

H

I

J

9. Continue finger cutting

After making that first cut, slip the kerf over the pin and cut another kerf (Figure **J**). This will create an alternating series of fingers and kerfs. Keep doing this for the entire end of the board, and repeat on your second test piece. It's OK if they don't line up as anticipated; I'll show you how to do that in a moment.

10. Test fitting

Test your cuts by fitting the two pieces together. They should interlock without much force (Figure **K**). If the fit is too loose you'll need to move the pin away from the blade; if it's too tight you'll need to move the pin toward the blade. This will change the size of the fingers. You will only need to move the pin over about the thickness of a business card. To do so, loosen your clamps slightly and nudge your board over. Tighten and repeat Steps 8 and 9.

Once you get that perfect fit, you can then screw your pin board to your sled or miter gauge. In this example I chose not to screw it in and just left the clamps on.

Everything look good? Great! Now it's time to make your box.

K

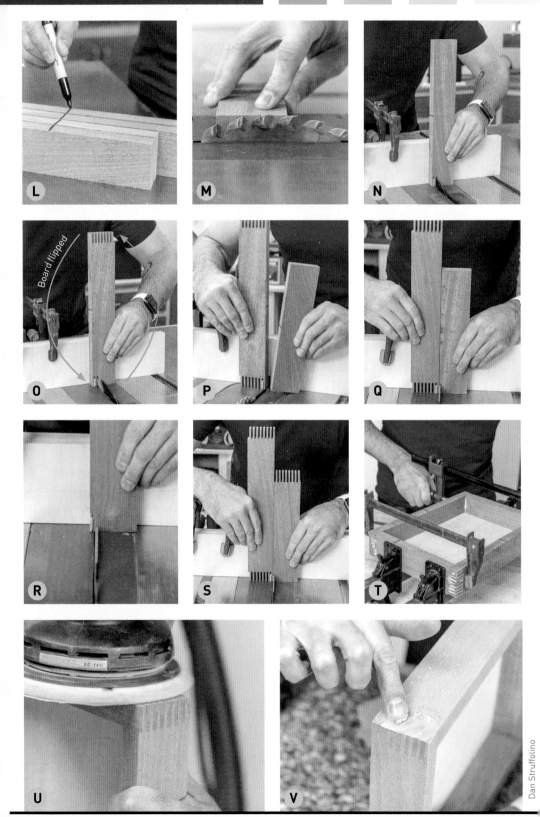

L

M

N

O

Board flipped

P

Q

R

S

T

U

V

Dan Struffolino

11. Setting up for success

First, mark the top of each of the four box pieces — you can line them up and make a mark all at once, like in Figure **L**. This mark is key to our success and will always face toward the pin when cutting the fingers.

12. Blade height

Set your table saw blade height to just a hair higher than the thickness of your boards (Figure **M**). This will give us fingers that are slightly too long — they'll extend off the side of the box, but we can sand them down after glue-up. It's better to have them too long than too short.

13. Long side finger cutting

We will start off with the two long pieces. Start making your fingers with the mark facing toward the pin (Figure **N**). Cut all the fingers just as you did with the test piece, along the entire end.

14. Flipped long side finger cutting

Flip the board over and cut the fingers on the other end (Figure **O**). Again, make sure the mark is facing inside, toward the pin. Once you're done, repeat Steps 13 and 14 on the other long board.

15. Short side registration

Now it's time to register the fingers for the short side. First, take one of the completed long sides and place it over the pin. Position the short side up against the long piece, once again making sure the marks on both pieces face inward, toward the pin, as seen in Figure **P**.

16. Start short side finger cutting

While holding both pieces tightly up against the fence you can run your board through the blade making that first cut in the short side (Figure **Q**).

17. Continue short side finger cutting

Now with that first kerf cut, you can continue to cut your fingers as in previous steps (Figure **R**).

18. Flipped short side finger cutting

Once you finish cutting the fingers on the short side end, you can flip the board over and cut the fingers on the opposite end as seen in Figure **S**. One more time, make sure the marks on both

pieces are facing inward toward the pin. Do I sound like a broken record yet? Those marks are key to a successful glue-up! After completing the fingers on the short piece, repeat Steps 15–18 on the remaining short piece.

19. Glue-up

Now it's time to glue it all together. If the joints were cut correctly your box will self-square. Make sure your marks are all facing up and everything should go together like a puzzle. Use clamps to keep it tight while the glue sets (Figure **T**).

20. Sanding flush

Once the glue dries you can sand the fingers flush with an orbital sander or disc sander (Figure **U**).

21. Filling gaps

Sometimes you may find tiny gaps after glue-up because of blow-out or fibers chipping away during the cut. You can easily cover this up by mixing up some sawdust and wood glue and filling the gap (Figure **V**). Once the glue dries you can sand everything flush again. This is a great trick for flawless-looking box joints.

Conclusion

That's all there is! I've made this jig several times and once you get the hang of it you can quickly build another jig in about 15 minutes. Even though I own one of the expensive adjustable store-bought jigs, I always reach for this shop-made version because it's already dialed in and ready to go.

And now that you know the technique, you can build all sorts of great wooden projects, from tool totes to drawers to heirloom boxes. ◗

To add a bottom to your box, there are several options. In this example I pre-cut grooves that will hold a piece of ⅛" plywood. You could always glue in a bottom after assembly, or even rout a rabbet after assembly and glue a bottom into that.

For more assistance, watch my video on this tutorial at makezine. com/go/table-saw-box-joints. I also have a version using only a router at makezine.com/go/router-box-joints.

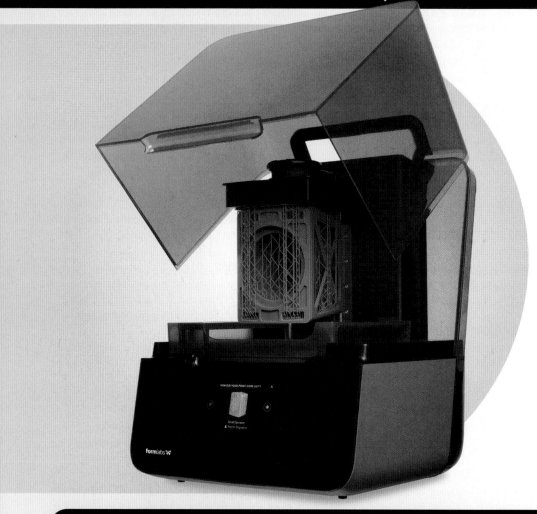

FORMLABS FORM 3 $3,499 formlabs.com

The folks at Formlabs are pretty smart. They know who their target audience is (people who are making things for a living) and they're leaning into that aspect. Instead of cutting away features to drop the price and trying to compete with the influx of dirt cheap machines, they've gone full steam ahead with the Form 3, trying to deliver the most feature-packed and advanced machine possible to people who have tight timelines on production work.

Formlabs is using a new system that they call LFS (low force stereolithography), with a new light path that they call the Light Processing Unit. They've decreased the size of the laser (increasing resolution), and added a ton of features to the software side of things. You can track all aspects of the printer and get notifications on your cellphone about your prints.

If I were building out a small makerspace, this may not be the resin printer I'd jump to first, but if I were equipping a small jewelry shop there'd be no question in my mind. I'd buy a Form 3.
—*Caleb Kraft*

Masked SLA printers ("MSLA") use an LCD to block and a strong UV light to cure the resin. They can be had for well under $500, and supply results that are not quite as good as a Form 3, but perfectly adequate for most hobbyists.

RODE WIRELESS GO

$278 rode.com

Rode's new Wireless Go microphone system starts to shake up the amateur audio world with a price, size, and feature set that are unlike anything else that's ever been available to hobbyists.

One of the most notable aspects of this kit is the transmitter, which is smaller than a box of Tic Tacs; it has the standard ⅛" micro-jack for a lav mic, but it also contains a built-in microphone of its own. Clip the transmitter on a shirt and get to work — no wires or additional mics needed. You can also plug in a shotgun mic for wireless boom jobs.

Rode states that this system is designed to work well in environments with lots of wireless signals, and using it for a full weekend at Maker Faire, I didn't find any electronic interference. But I do find that it's best when there's a clear path between the transmitter and receiver — going behind a door or even having the subject turn away from the transmitter at a moderate distance caused dropouts in my tests. But for the size, price, and versatility, this is a great little package for budding YouTubers, project documenters, and filmmakers.
—*Mike Senese*

NIKON Z6

$1,996 body-only
$2,596 with 24-70mm f/4 Z-mount lens
nikon.com

For any photography buff, the newest mirrorless cameras offer fascinating features, especially for video creators. Over the last few months I've been testing out the new Nikon Z6 camera to see what these advances can do. Takeaway: I'm convinced.

The electronic viewfinder was initially one of my biggest hesitations; it took me a couple days to adjust to it, but now it feels natural. I really like the eye-tracking focus (and focus peaking), the built-in stabilization, and the ability to instantly preview camera settings. I found myself getting more usable shots than from my full-frame DSLR.

I didn't love having to use pricey XQD cards (I've got drawers of cheap SD cards, none of which have ever failed me on a shoot). And while Nikon's new Z-mount 24-70mm f/4 kit lens focuses quickly and quietly, and takes clean shots from edge to edge, I've largely stuck with my F-mount 24-70mm f/2.8 with their FTZ adapter. This also works great with the old lenses I've collected through the years. For me, that's the best — marrying new technology with that from decades ago, and having it work brilliantly. —*Mike Senese*

TOOLBOX

DJI OSMO POCKET

$399 dji.com

When it comes to filming your DIY builds, DJI has been coming up with some cool tools. Typically known for their drones, they've started putting some of their gimbal technology into handheld cameras. The Osmo Pocket is a high resolution camera and gimbal that fits in your pocket.

Shooting 4k at 60fps is standard, resulting in stunning footage. Several other features set this apart as well, such as the ability to physically track your face (fantastic if you're filming alone while explaining a process), and the ability to take panning and tilting timelapse video out of the box.

Osmo Pocket's handle contains a tiny preview screen, but adjusting settings with it can be difficult if you've got big fingers. However, it also attaches to your phone for even more control and a larger display.

This is a direct competitor to typical box-style action cameras since it falls in the same price range and has similar image sensor stats, but the added features from the gimbal really make it stand out. They even have a waterproof case for diving, which might be nice to keep sawdust from my shop getting into the mechanical parts. —*Caleb Kraft*

FROM THE BOARD GUIDE

ADAFRUIT PYRULER
$12 adafruit.com

The PyRuler is just what the name says it is: A 6" ruler that is also a microcontroller ready to be programmed in Python. It not just measures centimeters and inches (and accurately at that — we checked!), but the back is covered in labeled pads so you can see where a surface mount electronic component fits, and read what that sort of chip is named. On top of that, it's also a miniature touch keyboard, because why not.

TEENSY 4.0
$20 pjrc.com

With a 600MHz clock speed, the Teensy 4.0 has horsepower beyond anything we're used to seeing in the microcontroller space. The NXP processor at the heart of Teensy 4.0 brings a smattering of other features uncommon in hobby microcontrollers: A CAN bus for communicating with automotive electronics, audio output, and processing for graphics and encryption like we'd normally expect to see on a single board computer rather than a microcontroller.

JACOBURGE TOUCHPAD 4.1
$72 jacoburge.co.uk

JacoBurges' "TouchPad" is a neat piece of kit that fully embraces the hackables moniker. Out of the box, it's a secondary keyboard made from a single circuit board, but its 6×6 grid of "keys" are actually programmable touch sensors. The notion is you can load the board up with hotkeys for your favorite program, scribble whatever icons you like on a clear plastic card that lays over the board, and use it as a productivity tool. —*Sam Brown*

Find more board reviews at makezine.com/comparison/boards

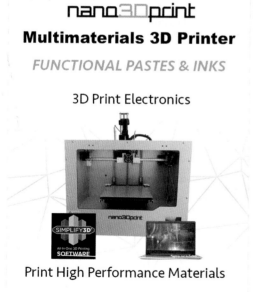

OVER THE TOP

Running the Numbers

Written by Florian Hu

This intricately geared 7-segment display gives a gorgeous view of your YouTube subscriber count

Recently my YouTube channel (youtube.com/huable) unexpectedly started to get attention, so I decided to do something special for when it finally reached 1,000 subscribers. An internet-connected live counter seemed like a great way to celebrate the event, with the added bonus that other people could build one themselves. There are already a few excellent split-flap kits out there so I came up with a mechanical 7-segment display.

I designed, printed, and tested a single segment first. It's a relatively simple concept. The segment needs to cycle through 10 states; while it's being actuated each state should be held for as long as possible before flipping over to the next. The states are encoded as knobs on timing gears, and these are different for each segment. A bi-stable spring helps with a crisp flip between states. Also, the segments "fold away" — instead of just rotating 90 degrees, they hinge around two pivots to hide the yellow display surface against the wall of the case. The entire CAD was done in Fusion 360, and everything can be printed with a standard FDM 3D printer.

Driving it with a single stepper motor appealed to me as an elegant, minimalistic solution. I also designed a custom circuit board to tidy everything up. The board is a minimal ATmega328, a Darlington driver array, and a Hall effect sensor, which drive the motor, register the current position after power up, and communicate with the next digit in the chain via serial. A Raspberry Pi gets the current count from the internet.

I also made each digit modular; when another digit is needed some time in the future, it can just be plugged in at the end. ⊘

Get the printable files: cults3d.com/en/3d-model/gadget/7-segments

FLORIAN HU works as a technical director in the VFX industry. He enjoys hobbies where he creates physical objects that do something in the real world.

Florian Hu